T0329667

Massive Connectivity

Massive Connectivity

Non-Orthogonal Multiple Access to High Performance
Random Access

Jinho Choi
Deakin University
Australia

IEEE PRESS

WILEY

Published by John Wiley & Sons, Inc., Hoboken, New Jersey.
Published simultaneously in Canada.

For general information on our other products and services or for technical support, please contact our Customer Care Department within the United States at (800) 762-2974, outside the United States at (317) 572-3993 or fax (317) 572-4002.

Wiley also publishes its books in a variety of electronic formats. Some content that appears in print may not be available in electronic formats. For more information about Wiley products, visit our web site at www.wiley.com.

Library of Congress Cataloging-in-Publication Data:

Names: Choi, Jinho, author.
Title: Massive connectivity : non-orthogonal multiple access to high
    performance random access / Jinho Choi.
Description: Hoboken, New Jersey : Wiley-IEEE Press, [2022] | Includes
    bibliographical references and index.
Identifiers: LCCN 2022015208 (print) | LCCN 2022015209 (ebook) | ISBN
    9781119772774 (cloth) | ISBN 9781119772781 (adobe pdf) | ISBN
    9781119772798 (epub)
Subjects: LCSH: Wireless communication systems.
Classification: LCC TK5103.2 .C456 2022 (print) | LCC TK5103.2 (ebook) |
    DDC 621.384–dc23/eng/20220609
LC record available at https://lccn.loc.gov/2022015208
LC ebook record available at https://lccn.loc.gov/2022015209

Cover Design: Wiley
Cover Image: © Fit Ztudio/Shutterstock

Set in 9.5/12.5pt STIXTwoText by Straive, Chennai, India

To my family

Contents

Author Biography

Jinho Choi was born in Seoul, Korea. He received B.E. (magna cum laude) degree in electronics engineering in 1989 from Sogang University, Seoul, and M.S.E. and Ph.D. degrees in electrical engineering from Korea Advanced Institute of Science and Technology (KAIST) in 1991 and 1994, respectively. He is with the School of Information Technology, Burwood, Deakin University, Australia, as a Professor. Prior to joining Deakin in 2018, he was with Swansea University, United Kingdom, as a Professor/Chair in Wireless, and Gwangju Institute of Science and Technology (GIST), Korea, as a Professor. His research interests include the Internet of Things (IoT), wireless communications, and statistical signal processing. He authored two books published by Cambridge University Press in 2006 and 2010. Prof. Choi received a number of best paper awards including the 1999 Best Paper Award for Signal Processing from EURASIP. He is on the list of World's Top 2% Scientists by Stanford University in 2020 and 2021. Currently, he is an Editor of *IEEE Wireless Communications Letters* and a Division Editor of *Journal of Communications and Networks* (JCN). He has also served as an Associate Editor or Editor of other journals including *IEEE Trans. Communications, IEEE Communications Letters, JCN, IEEE Trans. Vehicular Technology*, and *ETRI* journal.

Preface

Wireless connectivity is an indispensable technology of our lives today. From smartphones to connected vehicles to remote controlled drones, many devices rely on wireless connectivity. In addition, there are a variety of wireless connectivity technologies including WiFi, Zigbee, and cellular systems. These technologies exist all around us and help us a lot in our daily life. They have been developed through various stages and will continue to evolve. Advances in wireless technology have resulted in a variety of new applications. For example, with the invention of radio, mankind began to quickly share information through radio news from the 1920s, and in the twenty-first century, most of mankind enjoys wireless Internet services.

Wireless communication utilizes the radio spectrum, which is part of the electromagnetic spectrum with frequencies between 30 Hz and 300 GHz, and most wireless systems including cellular use microwave bands (1–100 GHz). For example, WiFi uses 2.4 and 5 GHz and fifth generation (5G) systems use sub-6 GHz frequency bands as well as millimeter bands (i.e. 24.25 GHz and above). In order to support an ever-increasing number of users and diverse applications, as technology advances, it is expected to increase the frequency so that a wider bandwidth is available. However, the bandwidth is a key limiting factor and scarce resource. Therefore, multiple access schemes to share a given bandwidth among users are always important to efficiently exploit the limited bandwidth. Orthogonal multiple access (OMA) schemes are currently employed for existing cellular systems. For example, the Global System for Mobile Communications (GSM) system, which is a second generation (2G) system, employs time division multiple access (TDMA) that allocates orthogonal time slots to different users. While OMA schemes are straightforward to be implemented and have been successfully employed in a number of wireless communication systems, there can be quite different approaches that can provide a higher spectral efficiency than them. Interestingly, such schemes had been discussed in the information theory literature since the 1970s under various names such as superposition coding, successive interference cancellation, and so on.

Code division multiple access (CDMA), which has been employed for a 2G system, i.e. Interim Standard 95 (IS-95), is a multiple access scheme for multiuser communications where multiple users coexist and share the same radio spectrum. In CDMA, coexisting users are differentiated by signature sequences that are not orthogonal. As a result, CDMA can be seen as a non-orthogonal multiple access (NOMA) and is expected to have higher spectral efficiency than an OMA scheme such as TDMA. In practice, in order to obtain high spectral efficiency in CDMA, an interference canceller, which was considered difficult to implement when IS-95 was introduced, is required together with precise power control.

Power-domain NOMA is another NOMA scheme where the power allocation is integrated with successive interference cancellation. Therefore, in power-domain NOMA, interference cancellation is essential in the receiver, and high spectral efficiency can be expected through this. This makes power-domain NOMA a strong multiple access technology candidate in the next generation cellular system.

Prior to the Internet-of-Things (IoT), it is not an exaggeration to say that most wireless connectivity technologies except for telematics were developed for human-type communication (HTC) services. As IoT applications become more popular, we are witnessing growth in machine-type communication (MTC) used in supporting the connectivity of numerous sensors and devices. MTC is now part of cellular systems such as 5G, and expected to play a more important role in next generation systems. As the number of things such as sensors and devices in MTC continues to increase, we expect to face various challenges due to limited spectrum. Like HTC, power-domain NOMA can be applied to MTC so that more devices can be supported with limited spectrum, which is the main topic of this book.

After discussing the well-known models for single-user and multiuser systems, this book discusses the details and key differences between OMA and NOMA. We then describe key principles of random access and extend random access to NOMA-based random access. We will also discuss how NOMA can be applied to MTC and take a closer look at how NOMA can improve the performance of MTC. After explaining NOMA from a communication perspective on their applications to MTC, at the end of this book, we will show how we can interpret these NOMA-based random access systems using game theory.

Although NOMA was not a new notion at all, I remember that there were skeptical views in the early studies of applying NOMA to wireless communication systems, which may be due to fear to the unknown or unconventional. I sincerely hope that these skeptical views will be turned into positive ones during the last decade of active research on NOMA, and would like to thank numerous researchers who have contributed to NOMA.

Finally, I would like to offer very special thanks to my wife, Kila, and two children, Seji and Wooji, for their understanding and love.

Melbourne, Australia
March, 2022

Jinho Choi

1

Introduction

Two main topics are covered in this book. One is machine-type communication (MTC) and the other is non-orthogonal multiple access (NOMA). Each topic has its own foundations and applications. In this chapter, we briefly explain each of them, and then explain why both topics should be covered in this book.

1.1 Machine-Type Communication

It may not be easy to imagine our daily life and business without the Internet although it began to appear as a backbone network in the 1970s to interconnect a small number of academic and military networks. The Internet is a network of networks and allows to exchange information between servers, computers, mobile phones, and so on restlessly. The Internet-of-Things (IoTs) is a natural extension of the Internet as machines, devices, and sensors are connected to the Internet to exchange information without human intervention in a number of applications such as smart factory.

As the number of devices connected to the Internet grows, their connectivity becomes important. Private and public networks can be used for their connectivity. For example, for smart home applications, a private network can be used at home to allow a small number of devices to be connected. For smart city applications, a large-scale public network would be preferable. Thus, the deployment of IoT networks depend on applications.

MTC is to support communications between machines or devices without human intervention. Unlike human-type communication (HTC), MTC mainly focuses on uplink transmission rather than downlink transmission (this is one of the main differences between MTC and HTC, where it can be seen that MTC's design principles must be different from those of HTC) and will support sporadic traffic in the form of short data packets. As a result, in order to keep signaling

Massive Connectivity: Non-Orthogonal Multiple Access to High Performance Random Access,
First Edition. Jinho Choi.
© 2022 The Institute of Electrical and Electronics Engineers, Inc. Published 2022 by John Wiley & Sons, Inc.

overhead low, the random access channel (RACH) procedure is used for MTC in Long-Term Evolution (LTE) systems. In the fifth generation (5G) system, a new random access scheme consisting of two steps, which is more efficient than the RACH procedure in LTE consisting of four steps, has been standardized.

MTC can provide connectivity for a large number of devices in a cell, paving the way for IoT applications to interact with devices deployed over a large area via cellular systems. This means that MTC becomes essential in various IoT applications such as smart cities and intelligent transport systems.

Furthermore, the global number of connected devices is expected to exceed 500 billion by 2030, while the human population is predicted to be 8.5 billion by the United Nations (UN). This means that IoT devices will outnumber human population by approximately 60-folds in 2030, and these devices will be used in a variety of IoT applications requiring heterogeneous connectivity demand. As a result, MTC will play a more prominent role in 5G and beyond (i.e. the sixth generation (6G)) and thus new MTC schemes need to be developed to meet the diverse requirements for future IoT applications.

1.2 Non-orthogonal Multiple Access

Various multiple access schemes have been used to support multiple users in a cellular system. For example, time division multiple access (TDMA) is adopted in the global system for mobile communications (GSM), which is regarded as a second generation (2G) system. In the third generation (3G) to 5G systems, orthogonal frequency division multiple access (OFDMA) is used. In general, most multiple access schemes used in cellular systems are orthogonal multiple access (OMA) schemes that allocate orthogonal channel resources to different users. To increase the spectral efficiency, NOMA schemes have been considered, where multiple users share the same channel resource.

NOMA has been extensively studied for cellular systems since the 2010s. In particular, for downlink transmissions, various NOMA schemes are studied using the difference in propagation loss between users near the center of the cell (where a base station (BS) is located) and users far from the center. The resulting NOMA is often referred to as power-domain NOMA.

It is necessary to distinguish between power-domain NOMA and NOMA in a broad sense. For example, code-division multiple access (CDMA) and interleave-division multiple access (IDMA) can be seen as NOMA schemes, because multiple users' signals can co-exist in a shared radio resource block, where one user signal can see the other users' signals as interfering signals. In CDMA, each user's signal is spread by a dedicated sequence, which is called the spreading sequence. Due to spreading sequences, CDMA has a bandwidth expansion. In particular, the bandwidth of CDMA increases by a factor of the processing gain or the length of the spreading sequence, while IDMA is a generalization of CDMA with forward error correcting codes. On the other hand, power-domain NOMA does not use spreading sequences. As a result, there is no bandwidth expansion and a high spectral efficiency can be achieved.

In power-domain NOMA, however, the transmit power levels and transmission rates should be carefully decided so that successive interference cancellation (SIC) can be used to remove other users' signals once they are decoded.

1.3 NOMA for MTC

In general, power-domain NOMA for downlink requires coordinated transmissions by a BS in terms of transmit powers and rates. Thus, it seems difficult to use power-domain NOMA for uplink as coordinated transmission by distributed users is not easy to implement. In other words, the gain of NOMA can be offset by excessive signaling overhead to perform coordinated transmissions by distributed users. As a result, the use of NOMA for random access seems quite challenging. On the contrary, NOMA is well-suited to random access as we will illustrate with two users.

Suppose that two users want to access a given channel without coordination. If two users always transmit their signals, they experience collisions and no user succeeds to transmit. Thus, they need to transmit with a certain probability. To this end, each user is to decide the access probability, denoted by p_k, $k = 1, 2$, for user k. The probability that at least one user succeeds to transmit a packet is given by

$$P_{\text{succ}} = p_1(1 - p_2) + p_2(1 - p_1).$$

If $p = p_k$, $P_{\text{succ}} = 2p(1 - p)$, which is maximized when $p = \frac{1}{2}$ and the maximum of P_{succ} is $\frac{1}{2}$, which is also the maximum average number of successfully transmitted packets. To consider random access with NOMA, we can assume two different power levels, P_H and P_L with $P_H > P_L$, and the receiver is able to decode both the signals if one user transmits a signal with transmit power P_H and the other with P_L. Let q be the probability that a user chooses the high transmit power when transmitting (with probability p). Then, the average number of successfully transmitted packets is given as follows:

$$\eta_2 = \underbrace{\binom{2}{1} p(1 - p)}_{\text{one user transmits}} + 2 \underbrace{\binom{2}{2} p^2}_{\text{two users transmit}} \binom{2}{1} q(1 - q)$$

$$= 2p(1 - p) + 4p^2 q(1 - q),$$

where the first term is the average number of successfully transmitted packets when only one user transmits and the second term is the average number of successfully transmitted packets when two users transmit simultaneously. It is easy to show that $q = \frac{1}{2}$ maximizes η_2. Then, we have $\eta_2 = 2p(1-p) + p^2 = 2p - p^2$. Thus, $p = 1$ maximizes η_2, which is 1. In other words, the maximum average number of successfully transmitted packets can be doubled if NOMA is used for uncoordinated transmissions of two users in uplink transmissions.

As more power levels are considered, the average number of successfully transmitted packets can increase. However, this comes at the cost of high transmit power by devices.

In this book, we mainly focus on the principles of NOMA and the application of NOMA to MTC. In particular, we discuss how NOMA can help improve the performance of random access in MTC once we present key ideas of random access including its stability. Game theory will also be used to understand the nature of random access where users compete for common radio resources in MTC.

1.4 An Overview of Probability and Random Processes

Prior to the main parts of this book, we present an overview of probability and random processes in this section, which can be used to see the required background in terms of probability and random processes. The reader is referred to well-known textbooks such as Papoulis and Pillai (2002); Ross (1995); Mitzenmacher and Upfal (2005) if not well equipped with theory of probability and random processes.

1.4.1 Review of Probability

A sample space Ω is the set of all possible outcomes (or events) of an experiment. Let A_m be an event m, which is a subset of Ω. A probability measure $\Pr(\cdot)$ is a mapping from Ω to the real line with the following properties:

1. $\Pr(A_m) \geq 0, A_m \in \Omega$
2. $\Pr(\Omega) = 1$
3. For a countable set of events, $\{A_m\}$, if $A_l \cup A_m = \emptyset$, for $l \neq m$, then

$$\Pr\left(\bigcup_{l=1}^{\infty} A_l\right) = \sum_{l=1}^{\infty} \Pr(A_l).$$

The joint probability of two events A and B is expressed as $\Pr(A \cap B) = \Pr(A|B)\Pr(B) = \Pr(B|A)\Pr(A)$, where the conditional probability of A given B is expressed as

$$Pr(A|B) = \frac{Pr(A \cap B)}{Pr(B)}, \quad Pr(B) > 0.$$

The two events A and B are independent if and only if

$$Pr(A \cap B) = Pr(A) Pr(B)$$

and this implies $Pr(A|B) = Pr(A)$.

In addition, for any two events A and B, we have

$$Pr(A \cup B) = Pr(A) + Pr(B) - Pr(A \cap B) \leq Pr(A) + Pr(B),$$

where the equality holds if $A \cap B = \emptyset$. Thus, for a set of events, it can be shown that

$$Pr\left(\bigcup_{l=1}^{\infty} A_l\right) \leq \sum_{l=1}^{\infty} Pr(A_l), \tag{1.1}$$

which is called the union bound.

1.4.2 Random Variables

A random variable is a mapping from an event ω in Ω to a real number, denoted by $X(\omega)$. We first consider continuous random variables. The *cumulative distribution function* (cdf) of X is defined as

$$F_X(x) = Pr(\{\omega | X(\omega) \leq x\})$$
$$= Pr(X \leq x)$$

and the *probability density function* (pdf) is defined as

$$f_X(x) = \frac{d}{dx} F_X(x),$$

where the subscript X on F and f identifies the random variable. If the random variable is obvious, the subscript is often omitted.

Note that we use capital letters to denote random variables in this book if necessary. For example, X is a random variable, while x is a variable. Random vectors will be written in boldface letters. For example, $\mathbf{x}$ is a random vector. This could lead to a confusion, because a vector (not a random vector) will also be written in boldface letters. To avoid this confusion, we will make clear indication if necessary.

There are some well-known pdfs of continuous random variables as follows:

1. Gaussian pdf with mean μ and variance σ^2:

$$f(x) = \frac{1}{\sqrt{2\pi\sigma^2}} \exp\left(-\frac{1}{2}\left(\frac{x-\mu}{\sigma}\right)^2\right).$$

As the Gaussian pdf is frequently used in this book, it is denoted by $\mathcal{N}(\mu, \sigma^2)$. If X is a Gaussian random variable with mean μ and variance σ^2, we write $X \sim \mathcal{N}(\mu, \sigma^2)$.

2. Exponential pdf $(a > 0)$:

$$f(x) = \begin{cases} a\, e^{-ax}, & \text{if } x \geq 0; \\ 0, & \text{otherwise.} \end{cases}$$

3. Rayleigh pdf $(b > 0)$:

$$f(x) = \begin{cases} \frac{x}{b}\, e^{-x^2/b}, & \text{if } x \geq 0; \\ 0, & \text{otherwise.} \end{cases}$$

4. Chi-square pdf of n degrees of freedom:

$$f(x) = \begin{cases} \frac{x^{(n-2)/2} e^{-x/2}}{2^{n/2} \Gamma(n/2)}, & \text{if } x \geq 0; \\ 0, & \text{otherwise,} \end{cases}$$

where $\Gamma(x)$ is the gamma function.

A joint cdf of random variables, $X_1, X_2, \ldots, X_n$, is expressed as

$$F_{X_1, X_2, \ldots, X_n}(x_1, x_2, \ldots, x_n) = \Pr\left(X_1 \leq x_1, X_2 \leq x_2, \ldots, X_n \leq x_n\right).$$

If the random variables are independent, the cdf is given by

$$F_{X_1, X_2, \ldots, X_n}(x_1, x_2, \ldots, x_n) = \prod_{i=1}^{n} F_{X_i}(x_i).$$

The joint pdf of random variables, $X_1, X_2, \ldots, X_n$, is then expressed as

$$f_{X_1, X_2, \ldots, X_n}(x_1, x_2, \ldots, x_n) = \frac{\partial^n}{\partial x_1 \partial x_2 \cdots \partial x_n} F_{X_1, X_2, \ldots, X_n}(x_1, x_2, \ldots, x_n).$$

For independent random variables, the joint pdf becomes a product of individual pdfs, i.e. $f_{X_1, X_2, \ldots, X_n}(x_1, x_2, \ldots, x_n) = \prod_{i=1}^{n} f_{X_i}(x_i)$. The conditional pdf of X_1 given $X_2 = x_2$ is

$$f_{X_1 | X_2}(x_1 | x_2) = \frac{f_{X_1 X_2}(x_1, x_2)}{f_{X_2}(x_2)}, \quad f_{X_2}(x_2) > 0.$$

Thus, if X_1 and X_2 are independent, it can be shown that

$$f_{X_1 | X_2}(x_1 | x_2) = \frac{f_{X_1 X_2}(x_1, x_2)}{f_{X_2}(x_2)} = f_{X_1}(x_1).$$

Let us consider the expectation and variance of X. First, the expectation of X is defined as

$$\mathbb{E}[X] = \int x f_X(x) \, dx.$$

In addition, the expectation of $g(X)$, a function of X, is

$$\mathbb{E}[g(X)] = \int g(x) f_X(x) dx.$$

The variance of X is defined as

$$\mathrm{Var}(X) = \mathbb{E}[(X - \mathbb{E}[X])^2]$$
$$= \int (x - \mathbb{E}[X])^2 f_X(x) dx.$$

The conditional mean of X given $Y = y$ is

$$\mathbb{E}[X|Y = y] = \int x f_{X|Y}(x|Y = y) dx.$$

The pdf of a real-valued joint Gaussian random vector, $\mathbf{x}$, is given as

$$f(\mathbf{x}) = \frac{1}{\sqrt{\det(2\pi\mathbf{C_x})}} \exp\left(-\frac{1}{2}(\mathbf{x} - \bar{\mathbf{x}})^\mathrm{T} \mathbf{C_x}^{-1}(\mathbf{x} - \bar{\mathbf{x}})\right),$$

where $\bar{\mathbf{x}} = \mathbb{E}[\mathbf{x}]$ and $\mathbf{C_x} = \mathbb{E}[(\mathbf{x} - \bar{\mathbf{x}})(\mathbf{x} - \bar{\mathbf{x}})^\mathrm{T}]$. Here, the superscript T stands for the transpose. For convenience, if $\mathbf{x}$ is a Gaussian random vector with mean $\bar{\mathbf{x}}$ and covariance matrix $\mathbf{C_x}$, we write $\mathbf{x} \sim \mathcal{N}(\bar{\mathbf{x}}, \mathbf{C_x})$.

If $\mathbf{x}$ is a circularly symmetric complex Gaussian (CSCG) random vector,

$$f(\mathbf{x}) = \frac{1}{\det(\pi\mathbf{C_x})} \exp\left(-(\mathbf{x} - \bar{\mathbf{x}})^\mathrm{H} \mathbf{C_x}^{-1}(\mathbf{x} - \bar{\mathbf{x}})\right),$$

where $\bar{\mathbf{x}} = \mathbb{E}[\mathbf{x}]$ and $\mathbf{C_x} = \mathbb{E}[(\mathbf{x} - \bar{\mathbf{x}})(\mathbf{x} - \bar{\mathbf{x}})^\mathrm{H}]$. For a CSCG random vector, we can show that

$$\mathbb{E}[(\mathbf{x} - \bar{\mathbf{x}})(\mathbf{x} - \bar{\mathbf{x}})^\mathrm{T}] = 0.$$

If $\mathbf{x}$ is a CSCG random vector with mean $\bar{\mathbf{x}}$ and covariance matrix $\mathbf{C_x}$, we write $\mathbf{x} \sim \mathcal{CN}(\bar{\mathbf{x}}, \mathbf{C_x})$.

While a continuous random variable X is real-valued (i.e. $X \in \mathbb{R}$), a discrete random variable X has a countable image. For example, if X can be either 0 or 1, it is a discrete random variable. The cdf of a discrete random variable X be defined as that of a continuous random variable. However, the pdf cannot be defined. Instead, the probability mass function (pmf), can be defined as

$$P_m = \mathrm{Pr}\,(X = x_m), \quad x_m \in \mathcal{X}, \tag{1.2}$$

where $\mathcal{X}$ represents of the image of X. The mean and variance of a discrete random variable are given by

$$\mathbb{E}[X] = \sum_{x \in \mathcal{X}} x\, \mathrm{Pr}\,(X = x) = \sum_m x_m P_m,$$

$$\mathrm{Var}(X) = \mathbb{E}[(X - \mathbb{E}[X])^2] = \sum_m (x_m - \mathbb{E}[X])^2 P_m. \tag{1.3}$$

There are a number of different discrete probability distributions. For example, the Bernoulli distribution is the discrete probability distribution of a random variable X that takes 1 with probability p and 0 with probability $1 - p$. The binomial distribution is the discrete probability distribution of the random variable that is the sum of the outcomes of n independent Bernoulli trials, which is given by

$$\Pr(X = i) = \binom{n}{i} p^i (1 - p)^{n-1}, \quad X \in \mathcal{X} = \{0, \dots, n\}. \tag{1.4}$$

In this case, we write $X \sim X \sim \mathrm{Bin}(n, p)$. Another important distribution is the Poisson distribution that is given by

$$\Pr(X = i) = \frac{\lambda^i}{i!} e^{-\lambda}, \quad X \in \mathcal{X} = \{0, \dots\}, \tag{1.5}$$

where $\lambda > 0$, and we write $X \sim \mathrm{Pois}(\lambda)$. It can be readily shown that $\mathbb{E}[X] = \lambda$ and $\mathrm{Var}(X) = \lambda$. If X follows the Poisson distribution with parameter λ, we write $X \sim \mathrm{Pois}(\lambda)$.

A sequence of random variables, X_n, $n = 1, 2, \dots$, is called an independent and identically distributed (iid) sequence if the X_n's are independent and their distribution are identical.

Let X_n be a sequence of independent random variables with $\bar{x}_n = \mathbb{E}[X_n]$ and $\sigma_n^2 = \mathrm{Var}(X_n)$. Consider the normalized sum as follows:

$$Y_k = \frac{S_k - \mathbb{E}[S_k]}{\mathrm{Var}(S_k)}$$

$$= \frac{\sum_{n=1}^{k} X_n - \bar{x}_n}{\sum_{n=1}^{k} \sigma_n^2},$$

where $S_k = \sum_{n=1}^{k} X_n$. If

$$\lim_{k \to \infty} \Pr(a \le Y_k \le b) = \frac{1}{\sqrt{2\pi}} \int_a^b e^{-\frac{y^2}{2}} \, dy \quad (a < b),$$

the sequence X_n, $n = 1, 2, \dots$, is said to satisfy the central limit theorem (CLT). If a sequence satisfies the following condition, called the Lyapunov condition:

$$\lim_{k \to \infty} \frac{\sum_{n=1}^{k} \mathbb{E}[|X_n - \bar{x}_n|^3]}{\left(\sum_{n=1}^{k} \sigma_n^2\right)^{\frac{3}{2}}} = 0,$$

this sequence satisfies the CLT. An example is a binary iid sequence, where $X_n \in \{0, 1\}$ and $\Pr(X_n = 1) = \frac{1}{2}$. In this case, $\mathbb{E}[|X_n - \bar{x}_n|^3] = 0$ and it satisfies the Lyapunov condition.

For a nonnegative random variable X and $a > 0$, it can be shown that

$$\Pr(X \geq a) \leq \frac{\mathbb{E}[X]}{a}, \tag{1.6}$$

which is called Markov's inequality. If X is not a nonnegative random variable, we can consider $Y = e^{\lambda X}$, which is a nonnegative random variable. Then, it can be shown that

$$\Pr(e^{\lambda X} \geq a) \leq \frac{\mathbb{E}[e^{\lambda X}]}{a} = \frac{M_X(\lambda)}{a}, \tag{1.7}$$

where $M_X(\lambda) = \mathbb{E}[e^{\lambda X}]$ is the moment generating function (MGF). Furthermore, let $a = e^{\lambda \tau}$. Then, we have

$$\Pr(X \geq \tau) \leq e^{-\lambda \tau} \mathbb{E}[e^{\lambda X}]. \tag{1.8}$$

A tight upper-bound on the tail probability $\Pr(X \geq \tau)$ can be obtained by minimizing the right-hand side (RHS) term with respect to $\lambda \geq 0$, i.e.

$$\Pr(X \geq \tau) \leq \min_{\lambda \geq 0} e^{-\lambda \tau} \mathbb{E}[e^{\lambda X}], \tag{1.9}$$

which is called the Chernoff bound.

1.4.3 Random Processes

A (discrete-time) random process is a sequence of random variables, $\{X_m\}$. The mean and autocorrelation function of $\{X_m\}$ are denoted by $\mathbb{E}[X_m]$ and $R_X(l, m) = \mathbb{E}[X_l X_m^*]$, respectively. A random process is called wide-sense stationary (WSS) if

$$\mathbb{E}[X_l] = \mu, \quad \forall l;$$
$$\mathbb{E}[X_l X_m^*] = \mathbb{E}[X_{l+p} X_{m+p}^*], \quad \forall l, m, p.$$

For a WSS random process, we have

$$R_X(l, m) = R_X(l - m).$$

The power spectrum of a WSS random process, $\{X_m\}$, is defined as

$$S_X(z) = \sum_{k=-\infty}^{\infty} z^{-k} R_X(k).$$

In addition, we can show that

$$R_X(k) = \frac{1}{2\pi} \int_{-\pi}^{\pi} S_X(e^{j\omega}) e^{jk\omega} \, d\omega,$$

where $j = \sqrt{-1}$. A zero-mean WSS random process is called white if

$$R_X(l) = \begin{cases} \sigma_x^2, & \text{if } l = 0; \\ 0, & \text{otherwise,} \end{cases}$$

where $\sigma_x^2 = \mathbb{E}[|X_l|^2]$.

If X_l is an output of the linear system whose impulse response is $\{h_p\}$ with a zero-mean white random process input, n_l, the autocorrelation function of $\{X_m\}$ is given by

$$R_X(l) = \sigma_n^2 \sum_l h_l h_{l-m},$$

where $\sigma_n^2 = \mathbb{E}[|n_l|^2]$. In addition, its power spectrum is given as

$$S_X(z) = \sigma_n^2 H(z) H(z^{-1})$$

or

$$S_X(e^{j\omega}) = \sigma_n^2 |H(e^{j\omega})|^2, \quad z = e^{j\omega}.$$

1.4.4 Markov Chains

In this subsection, we briefly present some key results of Markov chains. The reader is referred to Norris (1998) for more details on Markov chains.

Suppose a discrete-time random process X_n for $n = 0, 1, \ldots$ takes on a finite or countable number of possible values from a set of nonnegative integers, i.e. $X_n \in \{0, 1, \ldots\}$. If $X_n = i$, it is said that the process is in state i at time n. A process has the Markov property if the state at time $n + 1$ depends only on the state at time n, i.e.

$$\Pr(X_{n+1} = j \mid X_n = i, X_{n-1} = i_{n-1}, \ldots, X_0 = i_0)$$
$$= \Pr(X_{n+1} = j \mid X_n = i),$$

and such a process is known as a Markov chain. Let us denote the transition probability from state i at time n to state j at time $n + 1$ by

$$\Pr(X_{n+1} = j \mid X_n = i) = p_{ij}.$$

It satisfies the following properties:

$$p_{ij} \geq 0 \quad \text{and} \quad \sum_j p_{ij} = 1.$$

To describe the transition probabilities for an arbitrary number of steps, let us consider the n-step transition probability as

$$p_{ij}^n = \Pr(X_{n+m} = j \mid X_m = i) \quad \text{for } n \geq 1. \tag{1.10}$$

As an example, let us consider a two-step transition probability as

$$p_{ij}^2 = \Pr(X_{2+m} = j \mid X_m = i) \tag{1.11}$$

$$= \Pr(X_2 = j \mid X_0 = i)$$

$$= \sum_k \Pr(X_2 = j \mid X_1 = k, X_0 = i)\Pr(X_1 = k \mid X_0 = i)$$

$$= \sum_k \Pr(X_2 = j \mid X_1 = k)\Pr(X_1 = k \mid X_0 = i)$$

$$= \sum_k p_{ik}p_{kj},$$

where the second line shows that the transition probabilities do not depend on time and Markov property has been used from the third line to the fourth line.

Let us consider the probability that the process goes from state i at time $t = 0$ and passes through state k at time $t = m$, and ends up at state j at time $t = n + m$. Similar to (1.11), this probability is obtained as

$$p_{ij}^{n+m} = \sum_k p_{ik}^n p_{kj}^m, \tag{1.12}$$

which is known as the Chapman–Kolmogorov equation.

Let us consider classes of states: State j is said to be accessible from state i if $p_{ij}^n > 0$ for some $n \geq 1$. It is said that two states i and j communicate if they are accessible to each other. This communication relation is denoted by $i \leftrightarrow j$. In addition, if two states communicate with each other, we say that two states belong to the same class. Accordingly, two different classes of states should be disjoint because they can communicate otherwise.

A Markov chain is said to be irreducible if all states communicate with each other.

If $P_{ii}^n = 0$ for all n that are not divisible by d, it is said that state i has period d. State i with $d = 1$ is said to be aperiodic. If two states i and j communicate with each other, the periods of the two states are the same.

Let ρ_{ij} denote the probability that the process makes a transition into state j given that the process starts in state i. State j is said to be recurrent if $\rho_{ij} = 1$ (for any i), and transient otherwise (i.e. $\rho_{ij} < 1$). Let ρ_{ij}^n represent the probability that the process makes the first transition into state j at time n since it starts in state i. Thus, $\rho_{ij} = \sum_{n=1}^{\infty} \rho_{ij}^n$. In addition, denote by μ_i the expected number of transitions to return to state i given that the process starts in state i. Then, for a recurrent state i, it can be shown that

$$\mu_i = \sum_{n=1}^{\infty} n\rho_{ii}^n. \tag{1.13}$$

State i is said to be positive recurrent if $\mu_i < \infty$, and null recurrent otherwise (i.e. $\mu_i = \infty$).

A positive recurrent and aperiodic state is called ergodic. For an irreducible Markov chain, if one state is ergodic, the other states are also ergodic. For an ergodic Markov chain, the stationary distribution, $\{\pi_i\}$, exists, which is defined as

$$\pi_j = \sum_{i=0}^{\infty} \pi_i P_{ij}. \tag{1.14}$$

In addition, for an ergodic Markov chain,

$$\pi_j = \lim_{n \to \infty} P_{ij}^n > 0. \tag{1.15}$$

2

Single-User and Multiuser Systems

In this chapter, we first look at the capacity of a single-user system. Then, we discuss two different types of multiuser systems such as point-to-multipoint and multipoint-to-point systems from information theoretic point of view. These multiuser systems are known as broadcast (downlink[1]) and multiple access (uplink) channels, respectively.

2.1 A Single-User System

Signal transmissions over wireless channels are limited by a number of factors. For example, due to propagation loss, the received signal power or signal-to-noise ratio (SNR) over a radio channel decreases with the distance between a transmitter and a receiver. In addition, due to fading and noise, the number of bits per unit time in signal transmissions becomes limited. Hence, it is important to understand the maximum transmission rate of a given channel, which is referred to as the achievable rate, in terms of a number of parameters. This section discusses the achievable rate and the notion of channel coding as a means of reliable signal transmissions.

2.1.1 Signal Representation

In digital communications, suppose that a data symbol, denoted by x, can be transmitted within a unit time. The unit time is inversely proportional to the system bandwidth, which is the bandwidth that is used by a communication system. In addition, x is an element of a signal constellation $\mathcal{X}$ that has a finite number of elements, i.e. $x \in \mathcal{X}$. For convenience, let M be the number of elements of $\mathcal{X}$, i.e. $M = |\mathcal{X}|$, and let $\mathcal{X} = \{a_1, \ldots, a_M\}$, where a_m represents the mth constellation point. In general, M is a power of 2. Thus, the number of bits that can be

---

1 In the context of cellular communications, downlink channels are the channels from a base station (BS) to users. On the other hand, uplink channels are the channels from users to a BS.

Massive Connectivity: Non-Orthogonal Multiple Access to High Performance Random Access,
First Edition. Jinho Choi.
© 2022 The Institute of Electrical and Electronics Engineers, Inc. Published 2022 by John Wiley & Sons, Inc.

transmitted by one data symbol becomes $\log_2 M$ and signal transmission with $\mathcal{X}$, where $|\mathcal{X}| = M$, is referred to as M-ary signaling.

Example 2.1   A signal to transmit through a given channel can have the in-phase and quadrature components. In particular, consider the following narrowband signal model:

$$A(t)\sin(2\pi f_c t + \theta(t)) = \underbrace{A(t)\cos(\theta(t))\sin(2\pi f_c t)}_{\text{in-phase}} + \underbrace{A(t)\sin(\theta(t))\cos(2\pi f_c t)}_{\text{quadrature}},$$

where f_c represents the carrier frequency, and $A(t)$ and $\theta(t)$ are slowly varying signals that bear information. Taking $\sin(2\pi f_c t)$ and $\cos(2\pi f_c t)$ as the two axes in the real and imaginary axes, respectively, the in-phase and quadrature components can be seen as complex variables in the complex plane. That is, a signal constellation point becomes $a_m = A_m \cos\theta_m + jA_m \sin\theta_m$, where $j = \sqrt{-1}$. In every symbol interval, information can be conveyed by sending different constellation points, i.e. $A(t)e^{j\theta(t)} \in \mathcal{X} = \{a_1, \ldots, a_M\}$, for $t = nT_s$, where T_s represents the symbol interval. Here, $A(t)e^{j\theta(t)}$ is referred to as the baseband signal, and the symbol rate, $\frac{1}{T_s}$, is much lower than the carrier frequency, i.e. $\frac{1}{T_s} \ll f_c$. The signal constellation for quadrature phase shift keying (QPSK) is a set of $M = 4$ complex variables, i.e. $\mathcal{X} = \left\{ \frac{+1+j}{\sqrt{2}}, \frac{+1-j}{\sqrt{2}}, \frac{-1+j}{\sqrt{2}}, \frac{-1-j}{\sqrt{2}} \right\}$, which are called constellation points. The complex plane is referred to as the signal space, and the four constellation points of QPSK can be shown in the signal space as in Figure 2.1. There should be a mapping between M constellation points and $\log_2 M$ bits, which can be done in a number of ways.

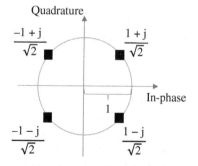

Figure 2.1   Signal space diagram for QPSK.

2.1.2   Transmission of Signal Sequences

We can generalize signal transmission to signal sequences based on the notion of channel coding (Lin and Costello, 1983, Wicker, 1995). Suppose that a signal sequence or vector of length L, $\mathbf{x} = [x_0 \cdots x_{L-1}]^T$, represents one of M messages. Thus, there should be M different signal vectors. That is, $\mathcal{X} = \{\mathbf{a}_1, \ldots, \mathbf{a}_M\}$ and $\mathbf{x} \in \mathcal{X}$, where $\mathcal{X}$ is now referred to as a codebook or code that is a set of signal vectors or codewords. Each codeword, $\mathbf{a}_m$, is a vector in an L-dimensional space. Note that since there are L elements in $\mathbf{x}$, the transmission of a codeword requires L unit times provided that each element can be transmitted over a unit time.

Let $\mathbf{y}$ be the received signal when $\mathbf{x}$ is transmitted through a given channel. For convenience, the transmitted and received signals are referred to as the channel input and output, respectively. It is often possible to find the relationship between the channel input and output. For example, suppose that the channel does not degrade the transmitted signal, while the receiver has the background noise that is modeled as a Gaussian random variable. This results in the additive white Gaussian noise (AWGN) channel, and the channel output is given by

$$y_l = x_l + n_l, \; l = 0, \ldots, L-1,$$

where n_l represents the noise at the receiver. Thus, we have

$$\mathbf{y} = \mathbf{x} + \mathbf{n}, \tag{2.1}$$

where $\mathbf{y} = [y_0 \cdots y_{L-1}]^T$ and $\mathbf{n} = [n_0 \cdots n_{L-1}]^T$. Here, the n_l's are independent zero-mean Gaussian random variables, i.e. $n_l \sim \mathcal{N}(0, \sigma^2)$, where σ^2 is the variance of noise. In (2.1), the signal power term is absorbed into x_l. Thus, we have $\mathbb{E}[|x_l|^2] = P$, where P stands for the signal power. Then, the SNR can be defined as

$$\text{SNR} = \frac{P}{\sigma^2}. \tag{2.2}$$

Example 2.2 Suppose that $x_l \in \{-\sqrt{P}, \sqrt{P}\}$. With $L = 3$, there can be $2^3 = 8$ signal vectors as shown in Figure 2.2. These vectors are equally spaced on a sphere of radius $\sqrt{P}$.

Figure 2.2 Signal space diagram for $M = 8$ signal vectors or codewords of length $L = 3$.

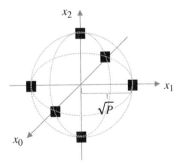

2.1.3 ML Decoding

In order to characterize the channel with randomness due to fading, interference, and background noise, we can consider the conditional distribution of **y** for given **x**, which is denoted by $f(\mathbf{y} \mid \mathbf{x})$. According to (2.1), $f(\mathbf{y} \mid \mathbf{x})$ for the AWGN channel is expressed as

$$
f(\mathbf{y} \mid \mathbf{x}) = f(y_0, y_1, \ldots, y_{L-1} \mid x_0, x_1, \ldots, x_{L-1})
$$

$$
= \prod_{l=0}^{L-1} f(y_l \mid x_l), \tag{2.3}
$$

where we have assumed that the y_l's are independent of each other for given x_l, and using $n_l \sim \mathcal{N}(0, \sigma^2)$ the conditional pdf $f(y_l \mid x_l)$ is expressed as

$$
f(y_l \mid x_l) = \frac{1}{\sqrt{2\pi\sigma^2}} \exp^{-\frac{(y_l - x_l)^2}{2\sigma^2}}.
$$

That is, for given x_l, y_l is a Gaussian random variable with mean x_l and variance σ^2. Thus, in (2.3), we have $f(\mathbf{y} \mid \mathbf{x}) \sim \mathcal{N}(\mathbf{x}, \mathbf{R_n})$, where $\mathbf{R_n} = \text{Cov}(\mathbf{n}) = \sigma^2 \mathbf{I}$.

While the conditional distribution, $f(\mathbf{y} \mid \mathbf{x})$, is used to characterize the channel, it can also be used for decoding. At a receiver, **y** is available as the channel output. However, the transmitted signal, **x**, is unknown and the receiver has to find it from **y**. In this case, **y** is given and **x** is a variable vector to be decided. That is, $f(\mathbf{y} \mid \mathbf{x})$ is a function of $\mathbf{x} \in \mathcal{X}$ for given **y**, which is now called the likelihood function.

Example 2.3 Suppose that $L = 1$ and $x = x_0 \in \{-\sqrt{P}, \sqrt{P}\}$. Then, for given $y = y_0$, the likelihood function is given by

$$
f(y \mid x) = \frac{1}{\sqrt{2\pi\sigma^2}} \exp\left(-\frac{(y-x)^2}{2\sigma^2}\right)
$$

$$
\in \left\{ \frac{1}{\sqrt{2\pi\sigma^2}} \exp\left(-\frac{(y+\sqrt{P})^2}{2\sigma^2}\right), \frac{1}{\sqrt{2\pi\sigma^2}} \exp\left(-\frac{(y-\sqrt{P})^2}{2\sigma^2}\right) \right\}.
$$

One decoding rule is to find the signal vector that maximizes the likelihood function $f(\mathbf{y} \mid \mathbf{x})$, which is called the maximum likelihood (ML) decoding. The ML solution is given by

$$
\hat{\mathbf{x}}_{\text{ml}} = \underset{\mathbf{x} \in \mathcal{X}}{\text{argmax}} \, f(\mathbf{y} \mid \mathbf{x}). \tag{2.4}
$$

For the AWGN channel, it can be shown that

$$
\hat{\mathbf{x}}_{\text{ml}} = \underset{\mathbf{x} \in \mathcal{X}}{\text{argmax}} \, f(\mathbf{y} \mid \mathbf{x})
$$

$$= \underset{x \in \mathcal{X}}{\operatorname{argmax}} \ \exp\left(-\frac{||\mathbf{y} - \mathbf{x}||^2}{2\sigma^2}\right)$$

$$= \underset{x \in \mathcal{X}}{\operatorname{argmin}} \ ||\mathbf{y} - \mathbf{x}||^2. \tag{2.5}$$

This implies that ML decoding for the AWGN channel is to find the closest codeword in $\mathcal{X}$ to the channel output $\mathbf{y}$ in terms of the Euclidean distance. The symbol error probability is given by

$$P_{\text{err}} = \Pr(\hat{\mathbf{x}}_{\text{ml}} \neq \mathbf{x}), \tag{2.6}$$

which is expected to be sufficiently low for reliable communications.

Example 2.4 Suppose that $L = 1$ and $x = x_0 \in \{-\sqrt{P}, \sqrt{P}\}$. For the AWGN channel, the ML decoding becomes

$$\hat{x} = \underset{x \in \mathcal{X}}{\operatorname{argmin}} |y - x|^2$$

$$= \begin{cases} \sqrt{P}, & \text{if } |y - \sqrt{P}|^2 < |y + \sqrt{P}|^2 \text{ or } y > 0, \\ -\sqrt{P}, & \text{otherwise} \end{cases}$$

If $x = -\sqrt{P}$, we have $y \sim \mathcal{N}(-\sqrt{P}, \sigma^2)$. Thus, the error probability when $x = -\sqrt{P}$ is $P_{\text{err}}|_{x=-\sqrt{P}} = \int_0^\infty \frac{1}{\sqrt{2\pi\sigma^2}} \exp\left(-\frac{(y+\sqrt{P})^2}{2\sigma^2}\right) dy = Q\left(\sqrt{\frac{P}{\sigma^2}}\right)$, where $Q(x) = \int_x^\infty \frac{1}{\sqrt{2\pi}} e^{-\frac{z^2}{2}} dz$ is the Q-function, which is the tail distribution function of the standard normal distribution and a decreasing function of x. Since $\text{SNR} = \frac{P}{\sigma^2}$ and the error probability when $x = \sqrt{P}$ is also $Q\left(\sqrt{\frac{P}{\sigma^2}}\right)$, we have $P_{\text{err}} = Q(\sqrt{\text{SNR}})$. Clearly, the error probability decreases with the SNR.

2.1.4 ML Decoding Over Fading Channels

In wireless communications, there can be multiple paths from a transmitter to a receiver depending on a given propagation environment. The resulting channel is called the multipath channel. Since each path may have a different propagation loss, the received signal at the receiver, denoted by $r(t)$, can be seen as a sum of weighted transmitted signals, i.e.

$$r(t) = \sum_{p=1}^{P} \alpha_p s(t - \tau_p) + n(t),$$

where $s(t)$ is the modulated signal, α_p is the channel attenuation of path p, P is the number of paths, and $n(t)$ is the background noise. Here, τ_p is the propagation delay of path p. For a linear modulation, we have $s(t) = x(t)e^{j2\pi f_c}$, where $x(t)$ and

$e^{j2\pi f_c}$ represent the baseband signal and the carrier, respectively. If the bandwidth of $x(t)$ is sufficiently narrow, it can be shown that

$$s(t - \tau_p) \approx x(t)e^{-j2\pi\tau_p} e^{j2\pi f_c}.$$ (2.7)

As a result, after demodulation and sampling, the received signal becomes

$$y_l = hx_l + n_l, \quad l = 0, \dots, L - 1,$$ (2.8)

where h represents the channel coefficient that is given by

$$h = \sum_{p=1}^{p} \alpha_p \, e^{-j2\pi\tau_p}.$$

The resulting channel in (2.8) can be classified as a time-invariant frequency-flat fading channel among four different classes of fading channels as follows:

- Time-invariant frequency-flat fading channels: It is the case when both the transmitter and receiver are stationary, the channel will not change or change slowly over time. In addition, the propagation delay is shorter than the symbol duration.
- Time-invariant frequency-selective fading channels: If the baud rate of the input signal is high enough that the propagation delay is longer than the symbol duration, the channel will exhibit frequency-selective gain due to inter-symbol interference (in this case, the approximation in (2.7) cannot be used).
- Time-varying frequency-flat fading channels: It is the case when the transmitter and/or receiver is moving and the symbol period is long enough that the propagation delay is longer than the symbol period. For a time-varying frequency-flat channel, h in (2.8) has to be $h(t)$ as it varies over time.
- Time-varying frequency-selective fading channels: These channels are also called doubly-selective fading channels, where the channel coefficients are time-varying and the time delays associated with the channel coefficients are longer than the symbol duration.

For a time-invariant frequency-flat fading channel as in (2.8), the likelihood function is given by

$$f(\mathbf{y} \mid \mathbf{x}, h) \propto \exp\left(-\frac{||\mathbf{y} - h\mathbf{x}||^2}{2\sigma^2}\right).$$ (2.9)

Thus, the ML solution is

$$\hat{\mathbf{x}}_{ml} = \underset{\mathbf{x} \in \mathcal{X}}{\operatorname{argmin}} ||\mathbf{y} - h\mathbf{x}||^2.$$ (2.10)

This requires to know the channel coefficient h. Thus, prior to signal transmissions, the transmitter can send a known signal, called pilot signal to allow the receiver to estimate the channel coefficient h. It is important to note that the SNR at the receiver depends on the channel coefficient. For example, if h can be absorbed into the input signal, the signal power becomes $\mathbb{E}[|hx_i|^2] = |h|^2 P$ and the receiver or output SNR becomes $\frac{|h|^2 P}{\sigma^2}$.

Example 2.5 Consider the case in Example 2.4 with (2.8). It can be shown that the error probability becomes $P_{\text{err}} = Q\left(\sqrt{\frac{|h|^2 P}{\sigma^2}}\right)$, which demonstrates that the error probability depends on the channel coefficient as well as the input SNR, $\frac{P}{\sigma^2}$, for fading channels.

2.1.5 Achievable Rate

From (2.5), it can be seen that a decoding error happens if there exists $\mathbf{x}' \in \mathcal{X}$ and $\mathbf{x} \neq \mathbf{x}'$ such as

$$||\mathbf{y} - \mathbf{x}'||^2 \leq ||\mathbf{y} - \mathbf{x}||^2.$$

Using (2.1), we have

$$||\mathbf{x} - \mathbf{x}' + \mathbf{n}||^2 \leq ||\mathbf{n}||^2. \tag{2.11}$$

It can be observed that the performance of ML decoding can be not only affected by the noise but also the distance between two codewords in $\mathcal{X}$. Accordingly, the minimum distance of two codewords in $\mathcal{X}$ is of particular importance to the decoding performance, which is defined as

$$d_{\mathcal{X}} = \min_{\mathbf{x}, \mathbf{x}' \in \mathcal{X},\ \mathbf{x} \neq \mathbf{x}'} ||\mathbf{x} - \mathbf{x}'||.$$

To guarantee a low (decoding) error probability, a codebook $\mathcal{X}$ should be designed to have a large minimum distance while codewords in $\mathcal{X}$ are uniformly distributed in an L-dimensional space. As the number M of elements increases, i.e. get crowded, it is expected that the minimum distance of code $\mathcal{X}$ decreases. Thus, for a low decoding error probability, the number of bits to send, i.e. $\log_2 M$, cannot be made arbitrarily large.

To see how large M can be in the AWGN channel, i.e. the number of bits transmitted reliably through the AWGN channel, let us make certain assumptions. First, we assume that $\mathbf{x}_i$'s are Gaussian vectors such that $\mathcal{X}$ has a set of M Gaussian vectors. Since each $\mathbf{x}_i \sim \mathcal{N}\left(0, P\mathbf{I}\right)$, codebook $\mathcal{X}$ can be called a Gaussian codebook with signal power P.

For a given $\mathbf{x} \in \mathcal{X}$ with a sufficiently large L, let us consider the radius of a sphere centered at $\mathbf{x}_i$:

$$||\mathbf{y} - \mathbf{x}_i|| = ||\mathbf{n}|| = \sqrt{\sum_l |n_l|^2} \to \sqrt{L\sigma^2}.$$

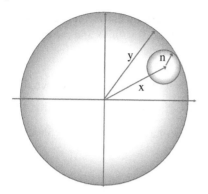

Figure 2.3 Sphere packing to find the capacity of AWGN channel.

On the other hand, since $y_l = x_l + n_l \sim \mathcal{N}(0, P + \sigma^2)$, the radius of **y** is expressed by

$$||\mathbf{y}|| = \sqrt{\sum_l |x_l + n_l|^2} \rightarrow \sqrt{L(P + \sigma^2)}. \tag{2.12}$$

As illustrated in Figure 2.3, the channel output, **y**, forms a big sphere of radius $\sqrt{L(P + \sigma^2)}$, while there are M small spheres centered at $\mathbf{x} \in \mathcal{X}$ and each has a radius of $\sqrt{L\sigma^2}$. The volume of L-dimensional sphere of radius r is $S_L r^L$, where S_L is a constant that only depends on L.

For a given **x**, the small sphere centered at **x** should not overlap any adjacent small spheres for a low decoding error probability. Thus, M small spheres can fill up the large spheres without overlapping each other (for a low decoding error probability), if

$$M S_L \left(\sqrt{L\sigma^2} \right)^L \leq S_L \left(\sqrt{L(P + \sigma^2)} \right)^L,$$

which can be rewritten as

$$M \leq \left(\frac{\sqrt{P + \sigma^2}}{\sqrt{\sigma^2}} \right)^L = \left(1 + \frac{P}{\sigma^2} \right)^{\frac{L}{2}}. \tag{2.13}$$

Let r_{code} denote the code rate, i.e. the number of bits per symbol or channel use, which can be expressed as $r_{\text{code}} = \frac{\log_2 M}{L}$. Using (2.13), r_{code} is given by

$$r_{\text{code}} = \frac{\log_2 M}{L} \leq \frac{1}{2} \log_2 \left(1 + \frac{P}{\sigma^2} \right), \tag{2.14}$$

where the upper-bound is the achievable rate or the channel capacity of the AWGN channel based on sphere packing. As shown in (2.14), the capacity is a function of SNR, i.e. $\frac{P}{\sigma^2}$, and an upper-bound on the transmission rate in bits per symbol. It is known that the probability of error in (2.6) decreases with L if the code rate r_{code} is less than the capacity (Cover and Thomas, 2006). Conversely, if r_{code} is higher than the capacity, the probability of error can be arbitrarily high. If complex-valued signals (with in-phase and quadrature-phase components) are transmitted, the capacity of the AWGN channel, which is called the Shannon capacity becomes doubled, i.e.

$$C_A = \log_2\left(1 + \frac{P}{\sigma^2}\right)$$
$$= \log_2\left(1 + \text{SNR}\right),\tag{2.15}$$

which is illustrated in Figure 2.4 as a function of SNR.

Consider the capacity of fading channels. When the channel output is given as in (2.8), the capacity conditioned on $|h|^2$ becomes

$$C_F(|h|^2) = \log_2\left(1 + \frac{|h|^2 P}{\sigma^2}\right).\tag{2.16}$$

The ergodic or average capacity is obtained by taking the mean of $C_F(|h|^2)$ over $|h|^2$. When $f_{\mathscr{H}}(v)$ denotes the pdf of channel gain $v \triangleq |h|^2$, the ergodic capacity can be obtained as

$$\bar{C} = \int_0^\infty C_F(v) f_{\mathscr{H}}(v) dv$$
$$= \int_0^\infty \log_2\left(1 + \frac{vP}{\sigma^2}\right) f_{\mathscr{H}}(v) dv.\tag{2.17}$$

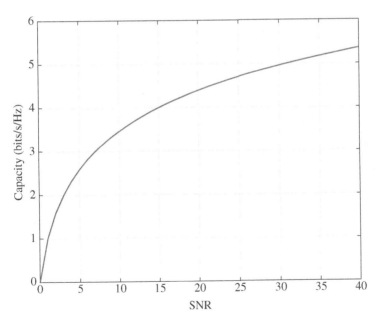

Figure 2.4 Shannon capacity of AWGN channel in terms of SNR.

Notice that to achieve the ergodic capacity in (2.17), the code rate r_{code} has to adapt to the channel gain $|h|^2$; that is, $r_{\text{code}} \leq C_F\left(|h|^2\right)$, when P is fixed. If r_{code} is fixed, successful transmissions depend on the random channel gain, $|h|^2$. The codewords cannot be decoded with the following (information outage) probability:

$$
\begin{aligned}
\mathbb{P}_{\text{out}} &= \text{Pr}\left(r_{\text{code}} > C_F(v)\right) \\
&= \text{Pr}\left(\log_2\left(1 + \frac{vP}{\sigma^2}\right) < r_{\text{code}}\right),
\end{aligned}
\tag{2.18}
$$

where (2.16) has been used. From (2.18), the ϵ-outage capacity is defined as

$$
C_\epsilon = \max_{\mathbb{P}_{\text{out}} \leq \epsilon} r_{\text{code}},
\tag{2.19}
$$

which is the highest code rate r_{code} such that the outage probability is less than or equal to ϵ.

Example 2.6 For frequency-flat fading, there are two typical fading channel models: Rayleigh and Rician fading channel models. The Rician fading channel model can be used when there is a line-of-sight (LoS) path. For the Rayleigh fading channel model, a channel without LoS is considered, where h is modeled as a circularly symmetric complex Gaussian (CSCG) random variable. Thus, the channel gain, v, is seen as the sum of the squares of two independent zero-mean Gaussian random variables with variance $\frac{\sigma_h^2}{2}$, which has the following exponential distribution:

$$
f_H(v) = \frac{1}{\sigma_h^2} \exp\left(-\frac{v}{\sigma_h^2}\right), \quad v \geq 0,
\tag{2.20}
$$

where $\sigma_h^2 = \mathbb{E}[|h|^2]$. This fading channel is called Rayleigh, because $\sqrt{v}$ follows a Rayleigh distribution, which is given by

$$
f(x = |h|) = \frac{2x}{\sigma_h^2} \exp\left(-\frac{x^2}{\sigma_h^2}\right), \quad x \geq 0.
\tag{2.21}
$$

From (2.20), the ϵ-outage capacity of Rayleigh fading channel is given by

$$
C_\epsilon = \log_2\left(1 + \ln\left(\frac{1}{1 - \epsilon}\right)\sigma_h^2 \text{SNR}\right).
$$

Note that with a probability of $1 - \epsilon$, the transmission at a rate of C_ϵ is possible. That is, since transmission is impossible with a probability of ϵ, the average transmission rate becomes $(1 - \epsilon)C_\epsilon$.

2.2 Multiuser Systems

If there are one transmitter and multiple receivers, the system has a point-to-multipoint communication, which is usually found in downlink channel. On the

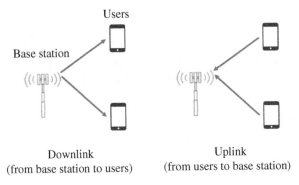

Users

Base station

Downlink
(from base station to users)

Uplink
(from users to base station)

Figure 2.5 Downlink and uplink transmissions in a cellular system consisting of a base station and multiple users.

other hand, if the system has multiple transmitters and one receiver, it is a multipoint-to-point communication, which takes place in uplink channel. As shown in Figure 2.5, the broadcast and multiple access channels are realized as downlink and uplink channels in cellular systems, respectively.

In a broadcast channel, the transmitter is a BS and the receivers are users, whereas the transmitters in the multiple access channel are users and the receiver the BS.

2.2.1 Broadcast Channels

In a broadcast channel, a BS broadcasts the information to users so that any one can receive it. In this section, we find the capacity of broadcast channel.

Let $\mathbf{s}$ be a signal vector of length L transmitted by the BS as a transmitter, i.e. $\mathbf{s} = [s_0 \ \cdots \ s_{L-1}]^T \in \mathbb{C}^L$, whereas $\mathbf{y}_k = [y_{k,0} \ \cdots \ y_{k,L-1}]^T \in \mathbb{C}^L$ for $k = 1, \ldots, K$ is the received vector of signal $\mathbf{s}$ at user k. Then, $\mathbf{y}_k$ is given by

$$\mathbf{y}_k = h_k \sqrt{P}\mathbf{s} + \mathbf{n}_k, \tag{2.22}$$

where h_k and P represent the channel coefficient from the BS to user k and the transmit power, respectively. In (2.22), $\mathbf{n}_k \sim \mathcal{CN}(0, N_0\mathbf{I})$ is the background Gaussian noise. We assume that $\mathbb{E}[s_l] = 0$ and $\mathbb{E}[|s_l|^2] = 1$ for all l's so that the transmit power per symbol is normalized to one.

In a broadcast channel, $\mathbf{s}$ contains messages to K users. To illustrate how each message to a user can be encoded into $\mathbf{s}$ using superposition coding, assume that $K = 2$, i.e. only two users. In addition, $\mathcal{S}_k$ and $\mathbf{s}_k$ denote the codebook for user k and a codeword in $\mathcal{S}_k$, respectively; that is, $\mathbf{s}_k \in \mathcal{S}_k$ for $k = 1, 2$. It is also assumed that $\mathbb{E}[\mathbf{s}_k] = 0$ and $\mathbb{E}[\mathbf{s}_k\mathbf{s}_k^H] = \mathbf{I}$, while $\mathcal{S}_k$ is assumed to be a Gaussian codebook such that $\mathbf{s}_k$ becomes a Gaussian vector. Using two codebooks, the BS encodes $\mathbf{s}$ in (2.22) for $\alpha \in (0, 1)$ as

$$\mathbf{s} = \sqrt{\alpha}\mathbf{s}_1 + \sqrt{1 - \alpha}\mathbf{s}_2. \tag{2.23}$$

We are interested in finding the transmission rate to user k, denoted by R_k. To find the expression for R_k, let divide both the sides of (2.22) by h_k for normalization purpose as follows:

$$\mathbf{z}_k = \frac{\mathbf{y}_k}{h_k} = \sqrt{P}\mathbf{s} + \bar{\mathbf{n}}_k, \tag{2.24}$$

where $\bar{\mathbf{n}}_k \sim C\mathcal{N}(0, \sigma_k^2 \mathbf{I})$ and $\sigma_k^2 = \frac{N_0}{|h_k|^2}$.

If $|h_1| > |h_2|$, which is equivalent to $\sigma_1^2 < \sigma_2^2$, the received signals at the two users are given by

$$\begin{aligned} \mathbf{z}_1 &= \sqrt{P}\mathbf{s} + \bar{\mathbf{n}}_1, \\ \mathbf{z}_2 &= \sqrt{P}\mathbf{s} + \bar{\mathbf{n}}_2 = \mathbf{z}_1 + \mathbf{w}, \end{aligned} \tag{2.25}$$

where $\mathbf{w} \sim C\mathcal{N}(0, \sigma_w^2 \mathbf{I})$ and $\sigma_w^2 = \sigma_2^2 - \sigma_1^2 > 0$.

Consider R_2 first. Substituting (2.23) into (2.25), the received signal at user 2 can be rewritten as

$$\mathbf{z}_2 = \sqrt{P(1 - \alpha)}\mathbf{s}_2 + \underbrace{\sqrt{P\alpha}\mathbf{s}_1 + \bar{\mathbf{n}}_2}_{=\text{interference+noise}}. \tag{2.26}$$

Since the term $\sqrt{P\alpha}\mathbf{s}_1 + \bar{\mathbf{n}}_2$ can be seen as a Gaussian noise with respect to user 2, the SNR of $\mathbf{z}_2$ becomes $\frac{P(1-\alpha)}{P\alpha + \sigma_2^2}$. The resulting achievable rate R_2 becomes

$$R_2 \leq \log_2 \left(1 + \frac{P(1 - \alpha)}{P\alpha + \sigma_2^2} \right). \tag{2.27}$$

To find R_1, by taking $\mathbf{s}_2$ as the first signal to be decoded at user 1, we have

$$\mathbf{z}_1 = \sqrt{P(1 - \alpha)}\mathbf{s}_2 + \underbrace{\sqrt{P\alpha}\mathbf{s}_1 + \bar{\mathbf{n}}_1}_{=\text{interference+noise}}. \tag{2.28}$$

The SNR of $\mathbf{z}_1$ becomes $\frac{P(1-\alpha)}{P\alpha + \sigma_1^2}$. Recall that $\sigma_1^2 < \sigma_2^2$ such that SNR of user 1 is higher than that of user 2, i.e. $\frac{P(1-\alpha)}{P\alpha + \sigma_1^2} > \frac{P(1-\alpha)}{P\alpha + \sigma_2^2}$. In (2.27), it can be found that user 1 can decode $\mathbf{s}_2$ first, although $\mathbf{s}_2$ is not the desired signal at user 1. Once user 1 decodes $\mathbf{s}_2$, $\mathbf{s}_2$ can be removed from $\mathbf{z}_1$. Then, user 1 can have the following signal:

$$\mathbf{z}_1 - \sqrt{P(1 - \alpha)}\mathbf{s}_2 = \sqrt{P\alpha}\mathbf{s}_1 + \bar{\mathbf{n}}_1. \tag{2.29}$$

In order for user 1 to decode $\mathbf{s}_1$ successfully, the following should hold for R_1:

$$R_1 \leq \log_2 \left(1 + \frac{P\alpha}{\sigma_1^2} \right). \tag{2.30}$$

From (2.27) and (2.30), the capacity region, i.e. a pair of R_1 and R_2 for $\alpha \in [0, 1]$ is characterized by

$$R_1 \leq \log_2\left(1 + \frac{\alpha P}{\sigma_1^2}\right) \quad \text{and} \quad R_2 \leq \log_2\left(1 + \frac{(1-\alpha)P}{\alpha P + \sigma_2^2}\right). \tag{2.31}$$

If $\alpha = 1$, it can be seen that

$$R_1 \leq \bar{R}_1 = \log_2\left(1 + \frac{P|h_1|^2}{N_0}\right) \quad \text{and} \quad R_2 = 0. \tag{2.32}$$

In this case, $\mathbf{s}$ delivers the message to user 1 only. On the other hand, if $\alpha = 0$, $\mathbf{s}$ delivers the message to user 2 only with $R_2 \leq \bar{R}_2 = \log_2\left(1 + \frac{P|h_2|^2}{N_0}\right)$. In summary, the coded signals in (2.23) for broadcast channels is called superposition coding, while removing a decoded signal as in (2.29) is referred to as successive interference cancellation (SIC).

Figure 2.6 depicts the capacity region in (2.31) as the shaded area. Notice that each point on the dashed line is a linear combination of $\bar{R}_1$ and $\bar{R}_2$, i.e. $(R_1, R_2) = (\beta \bar{R}_1, (1 - \beta)\bar{R}_2)$, where $\beta \in [0, 1]$.

A fraction of the shared channel is exclusively allocated to user 1 (or user 2) as much as β (or $1 - \beta$). The resulting allocation is referred to as orthogonal allocation. It can be observed that superposition coding allows each user to have a higher achievable rate than orthogonal allocation.

2.2.2 Multiple Access Channels

In a multiple access channel, $\mathbf{s}_k$ represents the signal vector of length L transmitted by user k. It is also a codeword from a Gaussian codebook with $\mathbb{E}[s_{k,l}] = 0$ and $\mathbb{E}[|s_{k,l}|^2] = 1$, where $s_{k,l}$ is the lth element of $\mathbf{s}_k$.

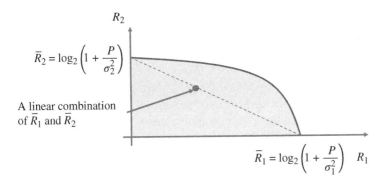

Figure 2.6 Capacity region for two-user broadcast channel.

When all the K users transmit simultaneously, the received signal $\mathbf{y}$ at the BS is given by

$$\mathbf{y} = \sum_{k=1}^{K} h_k \sqrt{P_k} \mathbf{s}_k + \mathbf{n}, \qquad (2.33)$$

where h_k and P_k are the channel coefficient from the kth user to the BS and the transmit power of the kth user, respectively, whereas $\mathbf{n} \sim \mathcal{CN}(0, N_0\mathbf{I})$ is the background Gaussian noise.

We assume $K = 2$, i.e. two users. When the receiver decodes the signal from user 1 first, it treats the signal from user 2 as an interfering signal. In this case, $\mathbf{y}$ can be written as

$$\mathbf{y} = h_1 \sqrt{P_1} \mathbf{s}_1 + \underbrace{h_2 \sqrt{P_2} \mathbf{s}_2 + \mathbf{n}}_{\text{interference+noise}}. \qquad (2.34)$$

Therefore, the achievable rate for user 1 becomes

$$R_1 \leq \log_2 \left(1 + \frac{P_1|h_1|^2}{P_2|h_2|^2 + N_0} \right). \qquad (2.35)$$

Provided that R_1 is less than the maximum achievable rate (the RHS), the BS can successfully decode user 1's signal, $\mathbf{s}_1$, and removes the corresponding component $h_1 \sqrt{P_1} \mathbf{s}_1$ from $\mathbf{y}$. The resulting signal is expressed as

$$\mathbf{y} - h_1 \sqrt{P_1} \mathbf{s}_1 = h_2 \sqrt{P_2} \mathbf{s}_2 + \mathbf{n}. \qquad (2.36)$$

This leads to the following achievable rate for user 2:

$$R_2 \leq \log_2 \left(1 + \frac{P_2|h_2|^2}{N_0} \right). \qquad (2.37)$$

A capacity region of two-user multiple access channel is then given by

$$
\begin{aligned}
R_1 &\leq \log_2 \left(1 + \frac{P_1|h_1|^2}{P_2|h_2|^2 + N_0} \right), \\
R_2 &\leq \log_2 \left(1 + \frac{P_2|h_2|^2}{N_0} \right).
\end{aligned}
\qquad (2.38)
$$

Figure 2.7 illustrates the capacity region of multiple access channel. The vertex A shows a pair of the maximum rates in (2.38). The vertex B shows a rate pair, if the signal from user 2 is decoded first. The maximum rates at B is given as follows:

$$
\begin{aligned}
R_1 &\leq \log_2 \left(1 + \frac{P_1|h_1|^2}{N_0} \right), \\
R_2 &\leq \log_2 \left(1 + \frac{P_2|h_2|^2}{P_1|h_1|^2 + N_0} \right).
\end{aligned}
\qquad (2.39)
$$

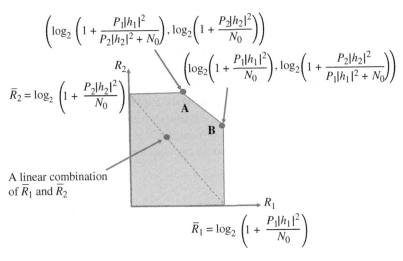

$$\left(\log_2 \left(1 + \frac{P_1|h_1|^2}{P_2|h_2|^2 + N_0} \right), \log_2 \left(1 + \frac{P_2|h_2|^2}{N_0} \right) \right)$$

$$\left(\log_2 \left(1 + \frac{P_1|h_1|^2}{N_0} \right), \log_2 \left(1 + \frac{P_2|h_2|^2}{P_1|h_1|^2 + N_0} \right) \right)$$

$$\bar{R}_2 = \log_2 \left(1 + \frac{P_2|h_2|^2}{N_0} \right)$$

A linear combination of $\bar{R}_1$ and $\bar{R}_2$

$$\bar{R}_1 = \log_2 \left(1 + \frac{P_1|h_1|^2}{N_0} \right)$$

Figure 2.7 Capacity region for two-user multiple access channel.

The capacity boundary between A and B can be obtained by taking a linear combination of A and B. The dashed line in Figure 2.7 is the set of two transmission rates for two users when the shared channel is split into two and assigned exclusively to two users according to orthogonal allocation. Each point on the dashed line is a linear combination of $\bar{R}_1 = \log_2 \left(1 + \frac{P_1|h_1|^2}{N_0} \right)$ and $\bar{R}_2 = \log_2 \left(1 + \frac{P_2|h_2|^2}{N_0} \right)$; that is, $(R_1, R_2) = (\beta\bar{R}_1, (1-\beta)\bar{R}_2)$ for $\beta \in [0,1]$. As mentioned before, β can be seen as the fraction of the shared channel that is exclusively allocated to user 1 (and $1 - \beta$ is the fraction of the shared channel that is exclusively allocated to user 2). Notice that as in Figure 2.7, orthogonal allocation in uplink communication is suboptimal.

Let us consider the sum rate $R_1 + R_2$ from (2.35) and (2.37), which is given by

$$R_1 + R_2 \le \log_2 \left(1 + \frac{P_1|h_1|^2}{P_2|h_2|^2 + N_0} \right) + \log_2 \left(1 + \frac{P_2|h_2|^2}{N_0} \right)$$

$$= \log_2 \left(1 + \frac{P_1|h_1|^2 + P_2|h_2|^2}{N_0} \right). \tag{2.40}$$

Regardless of the decoding order, the sum rate with K users can be given by

$$\sum_{k=1}^{K} R_k \le \log_2 \left(1 + \frac{\sum_{k=1}^{K} P_k|h_k|^2}{N_0} \right). \tag{2.41}$$

The result in (2.41) can also be obtained directly from (2.33). That is, since $\mathbf{s}_k$ is assumed to be a Gaussian signal vector, the sum of K signals, i.e. $\sum_{k=1}^{K} h_k \sqrt{P_k}\mathbf{s}_k$, is also Gaussian. If it is regarded as the desired signal, the capacity is given in the RHS of (2.41). This implies that the BS has to decode jointly all K signals, but it requires a prohibitively high computational complexity. For example, suppose that the size of codebooks for $\mathbf{s}_k$ is M_k. Then, for joint decoding, the BS has to find an optimal solution in the space of $\prod_{i=1}^{K} M_i$ elements. With $K = 2$ and $M_k = 2^{10} = 1024$, the size of the signal space becomes $M^2 = 2^{20} \approx 1.04 \times 10^6$. However, by exploiting superposition coding, it is possible to decode each signal at a time with SIC as described above. In this case, the total complexity of decoding becomes $\sum_{k=1}^{K} M_k$. Thus, if $K = 2$ and $M_k = 1024$, the size of the signal space becomes $2M = 2048$. This demonstrates that superposition coding also provides an efficient way to decode K signals that are transmitted simultaneously for uplink communication.

2.3   Further Reading

Fundamentals of digital communication are well covered by a number of textbooks, e.g. Proakis (2000); and Gallager (2008). For more details of wireless communications, the reader is referred to Biglieri (2005) and Tse and Viswanath (2005). There are also a number of textbooks on information theory, e.g. Cover and Thomas (2006).

3

OMA and NOMA

In this chapter, based on the notion of the capacity of broadcast and multiple access channels in Chapter 2, we briefly examine the capacity of orthogonal multiple access (OMA) schemes. Then, we introduce downlink and uplink non-orthogonal multiple access (NOMA) systems and discuss their fundamental aspects. We also discuss the power and rate allocation for NOMA. Finally, code division multiple access (CDMA) is introduced, which can also be seen as a NOMA scheme.

3.1 Orthogonal Multiple Access

This section examines well-known OMA schemes that are used in cellular systems as well as other wireless systems such as WiFi.

3.1.1 Time Division Multiple Access

Suppose that the system has K users and the length of time frame is T_{frame} (sec). For time division multiple access (TDMA), the frame is equally divided into K slots so that the length of a slot is $T_{\text{slot}} = \frac{T_{\text{frame}}}{K}$. Each slot is assigned to each user as illustrated in Figure 3.1. Since a slot can be seen as an orthogonal resource, TDMA is an OMA scheme.

Let T_{sym} be the symbol rate (in symbols per second). Then, each slot can carry $N_{\text{slot}} = \frac{T_{\text{slot}}}{T_{\text{sym}}}$ symbols. As a result, each user has a transmission rate of $\frac{N_{\text{slot}}}{T_{\text{frame}}} = \frac{1}{KT_{\text{sym}}} = \frac{r_{\text{sys}}}{T_{\text{sym}}}$ in symbols per second, where $r_{\text{sys}} = \frac{1}{T_{\text{sym}}}$ (symbols per second) is the system transmission rate in symbols per second. The system transmission rate is decided by the system bandwidth. Thus, for a given system bandwidth, each user's transmission rate decreases with the number K of users in the TDMA system.

Massive Connectivity: Non-Orthogonal Multiple Access to High Performance Random Access,
First Edition. Jinho Choi.
© 2022 The Institute of Electrical and Electronics Engineers, Inc. Published 2022 by John Wiley & Sons, Inc.

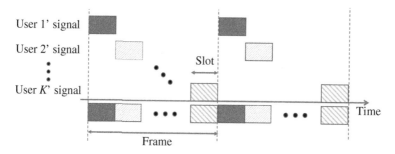

Figure 3.1 An illustration of TDMA with K users.

In finding the achievable rate, we assume that the system bandwidth is normalized to 1. If $K = 1$, the maximum achievable rate over an additive white Gaussian noise (AWGN) channel becomes $\log_2\left(1 + \frac{P_1}{N_0}\right)$. For $K \geq 2$, as mentioned above, each user can only use a slot per frame. Thus, user $k \in \{1, 2, \ldots, K\}$ has the following achievable rate:

$$R_k \leq \frac{1}{K}\log_2\left(1 + \frac{P_k}{N_0}\right). \tag{3.1}$$

TDMA can be employed for both uplink and downlink transmissions. In downlink transmissions, the achievable rate for user k is bounded as

$$R_k \leq \frac{1}{K}\log_2\left(1 + \frac{P|h_k|^2}{N_0}\right) = \frac{1}{K}\log_2\left(1 + \frac{P}{\sigma_k^2}\right), \tag{3.2}$$

where h_k represents the channel coefficient from user k to the base station (BS) and $\sigma_k^2 = \frac{N_0}{|h_k|^2}$. For uplink transmissions, the achievable rate for user k is bounded as

$$R_k \leq \frac{1}{K}\log_2\left(1 + \frac{P_k|h_k|^2}{N_0}\right) = \frac{1}{K}\log_2\left(1 + \frac{P_k}{\sigma_k^2}\right), \tag{3.3}$$

where h_k represents the channel coefficient from the BS to user k and P_k is the transmit power of user k. In Figure 3.2, the achievable rates of TDMA for downlink and downlink are illustrated when $K = 2$. It can be seen that TDMA is unable to achieve the capacity of broadcast channel for downlink transmissions as well as that of multiple access channel for uplink transmissions.

3.1.2 Frequency Division Multiple Access

In frequency division multiple access (FDMA), a system bandwidth, denoted by B, is divided into multiple bands in the frequency domain and each bandwidth is

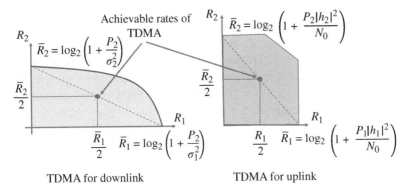

TDMA for downlink TDMA for uplink

Figure 3.2 Achievable rates of two-user TDMA in downlink and uplink.

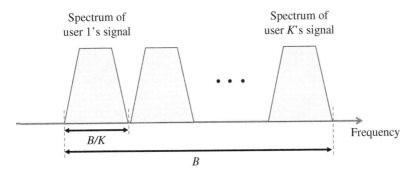

Figure 3.3 An illustration of FDMA with K users.

allocated to a user as illustrated in Figure 3.3. If there are K users, each user can have a bandwidth of $\frac{B}{K}$.

In a single-user system over the AWGN channel, user 1 with transmit power P can exploit a full system bandwidth, and has the achievable rate that is upper-bounded as

$$R_1 \leq \log_2\left(1 + \frac{P}{N_0}\right). \tag{3.4}$$

With the system bandwidth, B, the actual transmission rate is upper-bounded by $B \log_2\left(1 + \frac{P}{N_0}\right)$. For $K \geq 2$, each user's transmission rate is reduced by K times, and the achievable rate of user k is upper-bounded as

$$R_k \leq \frac{1}{K}\log_2\left(1 + \frac{P_k}{N_0}\right). \tag{3.5}$$

Notice that the achievable rates of downlink and uplink in (3.4) and (3.5) are identical to (3.2) and (3.3), respectively. This means that TDMA and FDMA, which are OMA schemes, have the same achievable rate.

Although both FDMA and TDMA are OMA schemes and have the same achievable rate, they have different features in terms of implementation. In realizing FDMA above, ideal bandpass filters are needed so that each user's signal cannot interfere with other users' signals in the adjacent bands. In practice, since such ideal bandpass filters cannot be implemented, it is necessary to include a guard band between two adjacent bands, which degrades the spectral efficiency.

On the other hand, TDMA requires synchronization between users as shown in Figure 3.1, whereas no synchronization is required in FDMA. For this reason, first generation (1G) cellular systems adopted FDMA using analog technology, while second generation (2G) cellular systems can be based on TDMA using advanced digital technology and provide a higher spectral efficiency.

3.1.3 Orthogonal Frequency Division Multiple Access

As another OMA scheme, orthogonal frequency division multiple access (OFDMA) is based on orthogonal frequency division multiplexing (OFDM) and can be seen as a generalization of OFDM to multiple users. It is expected that OFDMA inherits the advantages and disadvantages of OFDM. The third generation (3G) and fourth generation (4G) cellular systems as well as WiFi adopted OFDMA.

In principle, OFDM is a multiplexing scheme to combine multiple signals into one signal to transmit over a physical channel. Let s_l denote the lth signal for $l = 0, \ldots, L-1$ and $e^{j\frac{2\pi nl}{L}} = \cos\left(\frac{2\pi nl}{L}\right) + j\sin\left(\frac{2\pi nl}{L}\right)$, where $j = \sqrt{-1}$, represents the lth subcarrier in OFDM with L subcarriers. Then, these L signals can be combined into another signal $\tilde{S}_n$ for $n = \{0, \ldots, L-1\}$ as follows:

$$\tilde{S}_n = \sum_{l=0}^{L-1} s_l \, e^{j\frac{2\pi nl}{L}}. \tag{3.6}$$

In (3.6), l and n can be seen as the frequency and time indices, respectively, while $\tilde{S}_n$ is a weighted sum of L subcarriers with weight s_l in the time domain. An OFDM transmitter transmits $\tilde{S}_n$ sequentially over time. Thus, to transmit $\tilde{S}_n$, a total of L symbol periods are required (provided that each symbol $\tilde{S}_n$ is transmitted over one symbol interval). With $L = 64$, a few subcarriers are shown in the time domain in Figure 3.4.

Letting $\tilde{\mathbf{s}} = [\tilde{S}_0 \ \cdots \ \tilde{S}_{L-1}]^T$ and $\mathbf{s} = [s_0 \ \cdots \ s_{L-1}]^T$, as a matrix-vector notation, (3.6) can be rewritten as

$$\tilde{\mathbf{s}} = \mathbf{F}^H \mathbf{s}, \tag{3.7}$$

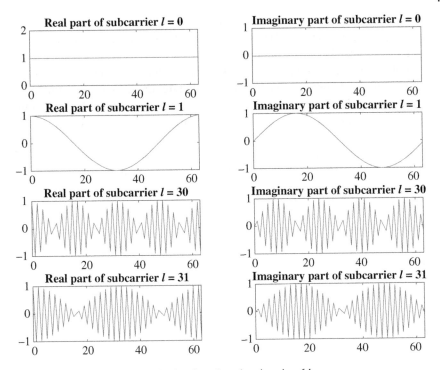

Figure 3.4 A few subcarriers in the time domain when $L = 64$.

where $\mathbf{F}$ is the discrete Fourier transform (DFT) matrix and the superscript H denotes Hermitian transpose. The (l, n)th element of $\mathbf{F}$ is $f_{l,n} = e^{-j\frac{2\pi nl}{L}}$. Notice that $\frac{1}{\sqrt{L}}\mathbf{F}$ is unitary, i.e.

$$\frac{1}{\sqrt{L}}\mathbf{F}\left(\frac{1}{\sqrt{L}}\mathbf{F}\right)^{H} = \left(\frac{1}{\sqrt{L}}\mathbf{F}\right)^{H}\frac{1}{\sqrt{L}}\mathbf{F} = \mathbf{I}.$$

Thus, the inverse of $\mathbf{F}$ is given by

$$\mathbf{F}^{-1} = \frac{1}{L}\mathbf{F}^{H}. \tag{3.8}$$

From this, $\tilde{\mathbf{s}}$ can be seen the output of inverse DFT of $\mathbf{s}$. To recover $\mathbf{s}$, the DFT can be applied to $\tilde{\mathbf{s}}$ as follows:

$$\mathbf{s} = \frac{1}{L}\mathbf{F}\tilde{\mathbf{s}} = \frac{1}{L}\mathbf{F}\mathbf{F}^{H}\mathbf{s} = \mathbf{s}. \tag{3.9}$$

In the context of OFDM, $\mathbf{s}$ is the signal vector in the frequency domain and $\tilde{\mathbf{s}}$ is the corresponding signal vector in the time domain. To transmit $\tilde{\mathbf{s}}$, which is referred to as OFDM symbol, a total of L symbol intervals are required in the time domain.

It seems OFDM merely transmits a transformed signal, $\tilde{\mathbf{s}}$, rather than $\mathbf{s}$ over L symbol periods. However, OFDM offers a number of advantages when the channel is frequency selective due to multipath.

As shown in (3.6), the lth signal s_l is to be transmitted over one OFDM symbol duration that is L times longer than the symbol duration. As a result, the bandwidth of each signal becomes L times narrower than that without OFDM. In the frequency domain, for each signal s_l, the channel becomes almost constant (or flat fading). For example, let H_l for $l = 0, \dots, L-1$ denote the channel coefficient corresponding to the lth subcarrier in the frequency domain. Then, the received signal by the lth subcarrier in the frequency domain is given by

$$y_l = H_l s_l + n_l, \tag{3.10}$$

where n_l represents the background white Gaussian noise. In (3.10), there is no inter-symbol interference although the signal is transmitted over a multipath channel.

For downlink communication, OFDM can be generalized to OFDMA with multiple users. Let $L = KM$, where K represents the number of users and M is the number of subcarriers per user. That is, $\mathbf{s}$ is a signal to K users as shown below:

$$\mathbf{s} = [\underbrace{s_0, \dots, s_{M-1}}_{\text{to user 1}}, \dots, \underbrace{s_{L-M}, \dots, s_{L-1}}_{\text{to user } K}]^{\mathrm{T}}. \tag{3.11}$$

Let $H_{k,l}$ denote the frequency domain channel gain for subcarrier l to user k. At user k, the received signal in the frequency domain is given by

$$y_{k,l} = H_{k,l} s_l, \ l = 0, \dots, L-1. \tag{3.12}$$

Then, among $\{y_{k,l}, \ l = 0, \dots, L-1\}$, user k only needs to detect signals through subcarriers $l \in \{(k-1)M, \dots, kM-1\}$. Clearly, there is no interference from the signals to the other users. This indicates that OFDMA is an OMA scheme.

Since multiple users' signals are transmitted through different sets of subcarriers, OFDMA is similar to FDMA. While FDMA needs small guard bands to separate the bandwidths allocated to users so as to prevent interference, which degrades the spectral efficiency, OFDMA does not require guard bands as all the subcarriers are orthogonal. This makes OFDMA more spectrum efficient than FDMA. Unfortunately, it is difficult to use OFDMA to uplink communication unless all the users can be perfectly synchronized. In general, the signaling overhead for controlling uplink OFDMA channels is overwhelming and can offset the improved spectral efficiency.

3.2 Non-Orthogonal Multiple Access

Unlike OMA, NOMA allows non-orthogonal allocation of a shared channel among multiple users as shown in Figure 3.5, which results in interference among them. Thus, interference mitigation is necessary in order to mitigate such interference. This section mainly focuses on NOMA of exploiting the difference of signal powers, which is often referred to as *power-domain* NOMA.

3.2.1 Downlink NOMA

As in broadcast channels, superposition coding plays a key role in downlink NOMA. In a two-user downlink system, suppose the superposition of two signals, $\mathbf{s}$, is given as in (2.23). Then, the received signal at user $k \in \{1,2\}$, denoted by $\mathbf{z}_k$, can be written as

$$\mathbf{z}_1 = h_1 \sqrt{P_1}\mathbf{s}_1 + h_1 \sqrt{P_2}\mathbf{s}_2 + \mathbf{n}_1,$$
$$\mathbf{z}_2 = h_2 \sqrt{P_2}\mathbf{s}_2 + h_2 \sqrt{P_1}\mathbf{s}_1 + \mathbf{n}_2, \tag{3.13}$$

where $P_1 = \alpha P$ and $P_2 = (1 - \alpha)P$, and α represents a fraction of total transmit power, P, to user 1. We assume that user 1 is close to the transmitter or BS, while user 2 is far away from the BS. This enables us to have $|h_1| > |h_2|$, because the propagation loss varies with distance. At user 2, i.e. the far user, it is expected that P_2 is sufficiently high so that the following signal-to-interference-plus-noise ratio (SINR) is reasonably high for signal decoding:

$$\text{SINR}_2 = \frac{|h_2|^2 P_2}{|h_2|^2 P_1 + N_0}. \tag{3.14}$$

Figure 3.5 Channel allocation for two users in multiple access schemes: (a) TDMA; (b) FDMA; (c) Power-domain NOMA.

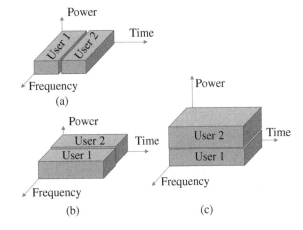

At user 1, i.e. the near user, the SINR becomes

$$\text{SINR}_1 = \frac{|h_1|^2 P_1}{|h_1|^2 P_2 + N_0}. \tag{3.15}$$

This shows that the increase of P_2 leads to the decrease of the SINR at user 1, which is not desirable for user 1. However, if P_2 is sufficiently high, user 1 can decode the signal to user 2, i.e. $\mathbf{s}_2$, before decoding $\mathbf{s}_1$. Let $\text{SINR}_{k';k}$ denote the SINR of user k when decoding the signal to user k' first. For convenience, let $\text{SINR}_{k;k} = \text{SINR}_k$. Suppose that user 1 decodes the signal to user 2, $\mathbf{s}_2$, first. In this case, the resulting SINR (taking $\mathbf{s}_2$ as the signal to be decoded at user 1) becomes

$$\text{SINR}_{2;1} = \frac{|h_1|^2 P_2}{|h_1|^2 P_1 + N_0}. \tag{3.16}$$

Provided that $P_2 > P_1$, it can be seen that $\text{SINR}_{2;1}$ could be sufficiently high for user 1 to decode the signal to user 2, $\mathbf{s}_2$, successfully. Once user 1 succeeds to decode $\mathbf{s}_2$, i.e. the interfering signal, successive interference cancellation (SIC) can be carried out to $\mathbf{z}_1$ in (3.13), which results in the following signal:

$$\mathbf{z}_1 - h_1 \sqrt{P_2} \mathbf{s}_2 = h_1 \sqrt{P_1} \mathbf{s}_1 + \mathbf{n}_1, \tag{3.17}$$

whose SNR becomes

$$\text{SNR}_1 = \frac{|h_1|^2 P_1}{N_0}. \tag{3.18}$$

If SNR_1 is sufficiently high, user 1 can succeed to decode the desired signal, $\mathbf{s}_1$.

Let Γ_i be the SINR threshold of signal $\mathbf{s}_i$ for successful decoding. Then, from (3.14), (3.16), and (3.18), for user 1 to successfully decode, the following inequalities must hold:

$$\begin{aligned} \text{SINR}_{2;1} &= \frac{|h_1|^2 P_2}{|h_1|^2 P_1 + N_0} \geq \Gamma_2, \\ \text{SNR}_1 &= \frac{|h_1|^2 P_1}{N_0} \geq \Gamma_1. \end{aligned} \tag{3.19}$$

For user 2, we should have

$$\text{SINR}_2 = \frac{|h_2|^2 P_2}{|h_2|^2 P_1 + N_0} \geq \Gamma_2. \tag{3.20}$$

From (3.19) and (3.20), it can be shown that if $|h_1| > |h_2|$,

$$\text{SINR}_2 = \frac{|h_2|^2 P_2}{|h_2|^2 P_1 + N_0} < \text{SINR}_{2;1} = \frac{|h_1|^2 P_2}{|h_1|^2 P_1 + N_0}.$$

Thus, if $|h_1| > |h_2|$, (3.19) and (3.20) are reduced to the following two inequalities to hold:

$$\text{SINR}_2 = \frac{|h_2|^2 P_2}{|h_2|^2 P_1 + N_0} \geq \Gamma_2,$$

$$\text{SNR}_1 = \frac{|h_1|^2 P_1}{N_0} \geq \Gamma_1. \tag{3.21}$$

Suppose that Gaussian codebooks are used for encoding to generate coded sequences or codewords, $\mathbf{s}_1$ and $\mathbf{s}_2$, and the transmission rate is decided according to the SINR threshold, i.e.

$$R_k = \log_2(1 + \Gamma_k).$$

Then, (3.21) becomes (2.31), which confirms that power-domain NOMA for downlink communication can provide a higher rate than OMA.

Power-domain NOMA for downlink communication can also be considered with a specific modulation scheme. For example, suppose that quadrature phase shift keying (QPSK) is employed, i.e. $s_k \in S$, where $S = \left\{ \frac{\pm 1 \pm j}{\sqrt{2}} \right\}$. The resulting signal can be expressed as

$$s = \sqrt{\alpha} s_1 + \sqrt{1 - \alpha} s_2. \tag{3.22}$$

This has a constellation of 16 points as shown in Figure 3.6.

For digital broadcast systems, a hierarchical modulation scheme based on the superposition of the signal constellations in (3.22) can be used to provide different service qualities depending on receiver's SNR. For example, s_1 in (3.22) can be the symbol that delivers the primary information bits, while s_2 is the symbol that delivers the secondary information bits. If the receiver has a sufficiently high SNR, it can decode both the primary and secondary information bits. However, if the SNR is not high enough, it can only decode the primary information bits. To this end, α

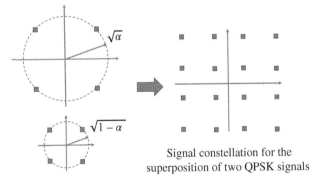

Signal constellation for the
superposition of two QPSK signals

Figure 3.6 A signal constellation of the superposition of two QPSK signals.

has to be reasonably high enough to do this. Although the purpose of hierarchical modulation, which is based on the superposition of multiple constellations, differs from that of downlink NOMA, it can be seen as a special case of downlink NOMA where the signal to user 2, say the far user, is also a useful signal to user 1, say the near user.

A generalization of power-domain NOMA in downlink with more than two users is relatively straightforward. Suppose K users and channel gains of these users are ordered as $|h_1|^2 > \cdots > |h_K|^2$. In addition, let P_k be the transmit power of the signal to user k and the total transmit power $P = \sum_{k=1}^{K} P_k$. Letting Γ_k be the threshold SINR at user k, for successful decoding at all K users, the SINR of user k for $k \in \{1, \ldots, K\}$ can be generalized as follows:

$$\text{SINR}_k = \frac{|h_k|^2 P_k}{|h_k|^2 \sum_{i=1}^{k-1} P_i + N_0} \geq \Gamma_k. \tag{3.23}$$

Especially, we have $\text{SNR}_1 = \frac{|h_1|^2 P_1}{N_0} \geq \Gamma_1$.

The minimum of transmit power P_k for $k \in \{1, \ldots, K\}$ can be recursively obtained. For example, the minimums of P_1 and P_2 should be

$$P_1^* = \frac{\Gamma_1 N_0}{|h_1|^2} \quad \text{and} \quad P_2^* = \frac{\Gamma_2(|h_2|^2 P_1^* + N_0)}{|h_2|^2}.$$

While the power allocation can be easily carried out as above, some difficulties may arise for the system with a large K. For example, user 1 in (3.23) is asked to decode all K signals, in order to decode own signal. Consequently, a large K becomes impractical.

There are also alternative approaches to support K users. Suppose $K = 2\overline{K}$ for a positive integer $\overline{K}$ and $\overline{K}$ orthogonal channel resource blocks. In each channel resource block, a pair of near and far users can be allocated with power-domain NOMA. This approach requires user pairing with $K = 2\overline{K}$ users and results in $\overline{K}$ two-user power-domain NOMA systems.

In addition, when the BS is equipped with multiple antennas, it is capable of forming multiple beams to target users. If a beam is orthogonal to each other, the signal transmitted through a beam to a user may not be interfering signals to the other users. Thus, using multiple beams, it is possible to support a number of users with a shared channel, which results in space division multiple access. With multiple beams, power-domain NOMA can be used to increase the number of users to be supported as in Higuchi and Kishiyama (2013) and Kim et al. (2013) through user pairing/clustering.

3.2.2 Uplink NOMA

It was seen that superposition coding has naturally arisen when all the users transmit simultaneously in uplink communication. Furthermore, as in (2.41), the sum rate was shown to be independent of decoding order. Thus, in order to maximize the sum rate, we can consider the power allocation without any specific decoding order.

The power allocation to maximize the sum rate is given by

$$\max_{\{P_k\}} \log_2\left(1 + \frac{\sum_{k=1}^{K} P_k |h_k|^2}{N_0}\right), \tag{3.24}$$

when K users access a shared channel for uplink transmission. If we add a constraint for a total transmit power, i.e. $\sum_{k=1}^{K} P_k \leq P$, where P is the total transmit power, the optimal allocation of maximizing the sum rate becomes

$$P_k = \begin{cases} P, & \text{if } |h_k|^2 > |h_q|^2, \ \forall q \neq k, \\ 0, & \text{else.} \end{cases} \tag{3.25}$$

The power allocation in (3.25) leads to a multiuser diversity system (Viswanath et al., 2002), where only one user with the highest channel gain at a time can access the shared uplink channel as illustrated in Figure 3.7. The resulting opportunistic access approach is optimal in terms of maximizing the sum rate subject to the total power constraint as shown above. Furthermore, when the channels are random due to the mobility of users, each user can be equally likely to have the highest channel gain. As a result, on average, each user has the same opportunity to access the shared channel.

On the other hand, if the channel is not time-varying or changes slowly, most users with low channel gain are starving for quite some time. This raises the issue of fairness. Thus, as an extreme case, let us consider uplink NOMA to support all the K users at an equal rate regardless of their channel gains. For convenience, we

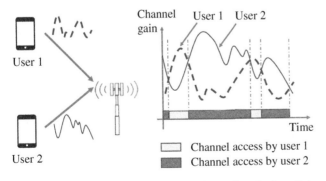

Figure 3.7 Opportunistic access for multiuser diversity in uplink communication with two users.

assume the descending decoding order with SIC. Then, the SINR are given by

$$SINR_K = \frac{|h_K|^2 P_K}{\sum_{q=1}^{K-1} |h_q|^2 P_q + N_0},$$

$$\vdots$$

$$SINR_k = \frac{|h_k|^2 P_K}{\sum_{q=1}^{k-1} |h_q|^2 P_q + N_0},$$

$$\vdots$$

$$SINR_1 = \frac{|h_1|^2 P_1}{N_0}. \tag{3.26}$$

In order to support all K users at an equal rate, the same SINR threshold, Γ, is applied to all the users. Thus, for example, when $K = 2$, we have

$$SINR_2 = \frac{|h_2|^2 P_2}{|h_1|^2 P_1 + N_0} = \Gamma,$$

$$SINR_1 = \frac{|h_1|^2 P_1}{N_0} = \Gamma. \tag{3.27}$$

Then, the transmit powers that satisfy (3.27) can be found as follows:

$$\frac{P_1}{N_0} = \frac{\Gamma}{|h_1|^2} \quad \text{and} \quad \frac{P_2}{N_0} = \frac{\Gamma(\Gamma + 1)}{|h_2|^2}. \tag{3.28}$$

For $K \geq 2$, for convenience, define

$$\Gamma_{(k)} = \Gamma \left(1 + \sum_{q=1}^{k-1} \Gamma_{(q-1)} \right), \tag{3.29}$$

where $\Gamma_{(0)} = 0$. Then, it is not difficult to show that

$$\frac{P_k}{N_0} = \frac{\Gamma_{(k)}}{|h_k|^2}, \quad k = 1, \dots, K. \tag{3.30}$$

Furthermore, letting $\Delta_k = \Gamma_{(k)} - \Gamma_{(k-1)}$, from (3.29) it can be shown that

$$\Delta_k = \Gamma_{(k)} - \Gamma_{(k-1)} = \Gamma \Gamma_{(k-1)}. \tag{3.31}$$

From this, it follows that

$$\Gamma_{(k)} = (1 + \Gamma)^{k-1} \Gamma, \quad k \geq 1. \tag{3.32}$$

As a result, for $\Gamma > 0$, $\Gamma_{(k)}$ increases with k. Thus, the descending decoding order becomes optimal in terms of minimizing the total transmit power, $\sum_{k=1}^{K} P_k$, if

$$|h_K|^2 \geq |h_{K-1}|^2 \geq \cdots \geq |h_1|^2 \tag{3.33}$$

according to the rearrangement inequality (Hardy et al., 1952). In other words, the decoding order at the BS is in order of high channel gain.

Example 3.1 Suppose that there are $K = 3$ users. For simplicity, let $N_0 = 1$. In addition, let $|h_1|^2 = 1$, $|h_2|^2 = 2$, and $|h_3|^2 = 10$. When $\Gamma = 2$, it can be shown that $\Gamma_{(1)} = 2$, $\Gamma_{(2)} = 6$, and $\Gamma_{(3)} = 18$. For the descending decoding order, from (3.30), we have $P_1 = \frac{2}{1} = 2$, $P_2 = \frac{6}{2} = 3$, and $P_3 = \frac{18}{10} = 1.8$. Thus, the total transmit power becomes 6.8, while other decoding orders result in higher total transmit powers.

3.3 Power and Rate Allocation

In this section, we briefly study the power and rate allocation for downlink NOMA by the BS. It is also possible to consider the power and rate allocation for uplink NOMA when users are expected to perform coordinated transmissions, although it is not discussed in this section.

3.3.1 System with Known Instantaneous CSI

In a digital communication system, a channel encoder is used for reliable transmission at a certain rate. For the receiver to decode the encoded signals successfully, the received SINR or SNR is expected to be higher than a threshold as illustrated in (3.19) and (3.20). To this end, the transmitter, i.e. the BS, carries out the transmit power allocation to meet the following inequalities:

$$P_1 \geq \frac{\Gamma_1 N_0}{|h_1|^2}, \tag{3.34}$$

$$P_2 \geq \frac{\Gamma_2 (\min(|h_1|^2, |h_2|^2) P_1 + N_0)}{\min(|h_1|^2, |h_2|^2)}, \tag{3.35}$$

where $\min(a, b)$ takes the minimum of a and b.

The BS wants to allocate P_1 and P_2 as small as possible. Let P_k^* denote the minimum of P_k in (3.34) and (3.34). Taking the equality in (3.34) and (3.34), P_k^* for $k \in \{1, 2\}$ becomes

$$P_1^* = \frac{\Gamma_1 N_0}{|h_1|^2} \quad \text{and} \quad P_2^* = \frac{\Gamma_2 \Gamma_1 N_0}{|h_1|^2} + \frac{\Gamma_2 N_0}{\min(|h_1|^2, |h_2|^2)}, \tag{3.36}$$

where we have plugged P_1^* into (3.35) for P_1 in order to get P_2^*. Then, the minimum total transmit power to meet the target SINRs is

$$P_t^* \triangleq P_1^* + P_2^*$$
$$= \frac{N_0 \Gamma_1 (1 + \Gamma_2)}{|h_1|^2} + \frac{N_0 \Gamma_2}{\min(|h_1|^2, |h_2|^2)}, \tag{3.37}$$

which is a function of $|h_1|^2$ and $|h_2|^2$. Therefore, when channel state information (CSI), i.e. h_1 and h_2, is available at the BS, the transmit power is controlled by the BS according to (3.37) to support two users with downlink NOMA. However, it can be seen that if one of the channel gain is too small due to deep fading, a very high transmit power is needed. This implies that reliable transmission is not always guaranteed under fading. In particular, for a transmit power budget P_{max}, the total transmit power can be controlled as follows:

$$P_1 + P_2 = \begin{cases} P_t^*, & \text{if } P_t^* \leq P_{max}, \\ 0, & \text{otherwise.} \end{cases} \tag{3.38}$$

This might be seen as a generalization of the truncated channel inversion policy. That is, if the total transmit power to meet the target SINRs is less than or equal to the maximum transmit power, P_{max}, the signals can be transmitted. Otherwise (i.e. the required total transmit power to meet the target SINRs exceeds the maximum transmit power, P_{max}), no signals are transmitted. In (3.38), if either $|h_1|^2$ or $|h_2|^2$ is sufficiently small, the BS is unable to meet the target SINRs for both the users and stops transmitting (because the users cannot decode signals), i.e. the total transmit power becomes 0, which results in an outage event.

Let us consider the outage probability that the BS is unable to support two users, which is denoted by $\mathbb{P}_{out}$. It can be defined as

$$\mathbb{P}_{out} = \Pr(P_1 + P_2 > P_{max}). \tag{3.39}$$

To find the expression of $\mathbb{P}_{out}$, we define the inverse of each term in (3.37) as

$$G_1 = \frac{|h_1|^2}{N_0 \Gamma_1 (1 + \Gamma_2)} \quad \text{and} \quad G_2 = \frac{\min\{|h_1|^2, |h_2|^2\}}{N_0 \Gamma_2}. \tag{3.40}$$

In addition, we define the two events as

$$A = \left\{ \frac{1}{G_1} + \frac{1}{G_2} > P_{max} \right\} \quad \text{and} \quad B = \{|h_1|^2 \geq |h_2|^2\}. \tag{3.41}$$

Then, the outage probability can be expressed as

$$\begin{aligned} \mathbb{P}_{out} &= \Pr(A) \\ &= \Pr(A \mid B^c) \Pr(B^c) + \Pr(A \mid B) \Pr(B). \end{aligned} \tag{3.42}$$

If the event B^c occurs, i.e. $|h_1|^2 < |h_2|^2$, the argument of event A in (3.41) is expressed as

$$\frac{1}{G_1} + \frac{1}{G_2} = \frac{N_0(\Gamma_1 + \Gamma_2 + \Gamma_1 \Gamma_2)}{|h_1|^2}, \tag{3.43}$$

which leads to

$$\Pr(A \mid B^c) = \Pr\left(|h_1|^2 < \frac{N_0(\Gamma_1 + \Gamma_2 + \Gamma_1 \Gamma_2)}{P_{max}} |h_1|^2 < |h_2|^2\right). \tag{3.44}$$

On the other hand, under the event of B, we have

$$\Pr(A \mid B) \Pr(B) = \Pr\left(\frac{1}{G_1} + \frac{1}{G_2} > P_{max} \mid |h_1|^2 \geq |h_2|^2\right) \Pr\left(|h_1|^2 \geq |h_2|^2\right)$$

$$= \text{Pr}\left(\frac{N_0\Gamma_1(1+\Gamma_2)}{|h_1|^2} + \frac{N_0\Gamma_2}{|h_2|^2} > P_{\max}, |h_1|^2 \geq |h_2|^2\right)$$

$$\leq \text{Pr}\left(\frac{N_0\Gamma_1(1+\Gamma_2)}{|h_1|^2} + \frac{N_0\Gamma_2}{|h_2|^2} > P_{\max}\right). \tag{3.45}$$

The inequality is due to the fact that $\text{Pr}(A, B) = \text{Pr}(A \cap B) \leq \text{Pr}(A)$. Consequently, the outage probability is upper-bounded as

$$\mathbb{P}_{\text{out}} \leq \underbrace{\text{Pr}\left(|h_1|^2 < \frac{N_0(\Gamma_1 + \Gamma_2 + \Gamma_1\Gamma_2)}{P_{\max}}\right) \text{Pr}(|h_1|^2 < |h_2|^2)}_{=(a)}$$

$$+ \underbrace{\text{Pr}\left(\frac{N_0\Gamma_1(1+\Gamma_2)}{|h_1|^2} + \frac{N_0\Gamma_2}{|h_2|^2} > P_{\max}\right)}_{=(b)}. \tag{3.46}$$

Under independent Rayleigh fading, $|h_k|^2$ becomes an exponential random variable with mean $\sigma_k^2 = \mathbb{E}[|h_k|^2]$ for $k \in \{1,2\}$ (the average power of channel gain). Under the assumption that user 1 is closer to the BS than user 2, it is expected that $\sigma_1^2 > \sigma_2^2$. Then, the first term in (3.46) is given by

$$(a) = \left(1 - \exp\left(-\frac{N_0(\Gamma_1 + \Gamma_2 + \Gamma_1\Gamma_2)}{\sigma_1^2 P_{\max}}\right)\right) \frac{\sigma_2^2}{\sigma_1^2 + \sigma_2^2}. \tag{3.47}$$

For the second term in (3.46), we need the following inequality:

$$\frac{1}{a} + \frac{1}{b} \leq 2\max\left(\frac{1}{a}, \frac{1}{b}\right) = \frac{2}{\min(a,b)} \quad \text{for} \quad a, b > 0.$$

From this, the term (b) in (3.46) is bounded as

$$(b) \leq \text{Pr}\left(\frac{2}{\min\left(\frac{|h_1|^2}{N_0\Gamma_1(1+\Gamma_2)}, \frac{|h_2|^2}{N_0\Gamma_2}\right)} > P_{\max}\right). \tag{3.48}$$

As the minimum of two independent exponential random variables is also an exponential random variable (Mitzenmacher and Upfal, 2005), the upper-bound in (3.48) can be expressed as

$$\text{Pr}\left(\frac{2}{\min\left(\frac{|h_1|^2}{N_0\Gamma_1(1+\Gamma_2)}, \frac{|h_2|^2}{N_0\Gamma_2}\right)} > P_{\max}\right) \leq 1 - \exp\left(-\frac{2N_0}{P_{\max}}\left(\frac{\Gamma_1(1+\Gamma_2)}{\sigma_1^2} + \frac{\Gamma_2}{\sigma_2^2}\right)\right).$$

$$\tag{3.49}$$

Finally, an upper-bound on the outage probability is given by

$$\mathbb{P}_{\text{out}} \leq \left(1 - \exp\left(-\frac{N_0(\Gamma_1 + \Gamma_2 + \Gamma_1\Gamma_2)}{\sigma_1^2 P_{\text{max}}}\right)\right) \frac{\sigma_2^2}{\sigma_1^2 + \sigma_2^2}$$
$$+ 1 - \exp\left(-\frac{2N_0}{P_{\text{max}}}\left(\frac{\Gamma_1(1 + \Gamma_2)}{\sigma_1^2} + \frac{\Gamma_2}{\sigma_2^2}\right)\right). \tag{3.50}$$

As shown in (3.50), for a low outage probability, it is important to keep the ratio $\frac{\sigma_1^2}{\Gamma_1(1+\Gamma_2)}$ high enough and make it comparable to $\frac{\sigma_2^2}{\Gamma_2}$. Thus, provided that $\Gamma_1 = \Gamma_2 = \Gamma$ (in this case, each user has the same transmission rate), σ_1^2 has to be $\Gamma + 1$ times greater than σ_2^2.

Figure 3.8 shows the outage probability and its upper-bound in (3.50) as functions of P_{max} when $\Gamma_1 = \Gamma_2 = 6$ dB and $\sigma_1^2 d^{-\eta} = \sigma_2^2$, where $d = 2$ and $\eta = 3$. This is the case that user 2 is two times farther away from the BS than user 1 and the path loss exponent is set to $\eta = 3$. It is assumed that $\sigma_2^2 = N_0$ so that P_{max} becomes the SNR at user 2.

Instead of using power control as above, the BS can control the transmission rates according to the channel conditions or CSI if channel encoders are flexible enough to adjust their code rates with fixed transmit powers, P_1 and P_2. To this end, the short-term rates are decided to be $\log_2\left(1 + \text{SNR}_1\right)$ and $\log_2\left(1 + \text{SINR}_2\right)$ for users 1 and 2, respectively, where SNR_1 and SINR_2 are given in (3.19) and

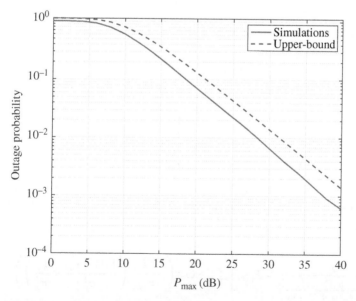

Figure 3.8 Outage probability (from simulations) and its upper-bound in (3.50).

(3.20), which are dependent on the channel power gains, $|h_1|^2$ and $|h_2|^2$. Since the channel power gains are varying, the long-term rates can be obtained by taking the expectation over the channel coefficients, i.e. the long-term rates are given by

$$r_1 = \mathbb{E}\left[\log_2\left(1 + \text{SNR}_1\right)\right] \quad \text{and} \quad r_2 = \mathbb{E}\left[\log_2\left(1 + \text{SINR}_2\right)\right]. \tag{3.51}$$

From (3.19), r_1 can be found as

$$r_1 = \int_0^\infty \log_2\left(1 + \frac{v_1 P_1}{N_0}\right) f_{H_1}(v_1) dv_1, \tag{3.52}$$

where $f_{H_1}(v_1)$ represents the pdf of $v_1 = |h_1|^2$. In addition, let $f_*(x)$ represent the pdf of $\min(|h_1|^2, |h_2|^2)$. Then, r_2 can be obtained as

$$
\begin{aligned}
r_2 &= \int_0^\infty \log_2\left(1 + \frac{xP_2}{xP_1 + N_0}\right) f_*(x) dx \\
&= \int_0^\infty \log_2\left(1 + \frac{x(P_1 + P_2)}{N_0}\right) f_*(x) dx - \int_0^\infty \log_2\left(1 + \frac{xP_1}{N_0}\right) f_*(x) dx.
\end{aligned}
\tag{3.53}
$$

Let us find the closed-form expressions for r_1 and r_2. When γ is the exponential random variable with unit mean, it can be shown that the (ergodic) capacity $C(\text{SNR})$ is Alouini and Goldsmith (1999)

$$C(\text{SNR}) = \mathbb{E}[\log_2(1 + \text{SNR}\gamma)]$$

$$= \frac{\exp\left(\frac{1}{\text{SNR}}\right)}{\ln 2} E_1\left(\frac{1}{\text{SNR}}\right), \tag{3.54}$$

where $E_1(x) = \int_x^\infty \frac{e^{-t}}{t} dt$ is the exponential integral. Thus, if $|h_k|^2$ is an independent exponential random variable (i.e. under independent Rayleigh fading), the long-term rates have the following closed-form expressions:

$$r_1 = C\left(\frac{\sigma_1^2 P_1}{N_0}\right),$$

$$r_2 = C\left(\frac{\sigma_1^2 \sigma_2^2 (P_1 + P_2)}{(\sigma_1^2 + \sigma_2^2) N_0}\right) - C\left(\frac{\sigma_1^2 \sigma_2^2 P_1}{(\sigma_1^2 + \sigma_2^2) N_0}\right). \tag{3.55}$$

Once P_1 and P_2 are given, the (long-term) rate allocation can be carried out for users 1 and 2 according to (3.55).

3.3.2 System with Unknown Instantaneous CSI

So far, we have assumed that the channel coefficients, h_k, are available at the BS. This is feasible if the users can send the BS their CSI, which may change slowly over time so that their variation during feedback delay is negligible. Here, the feedback delay is the time difference between the time when a user estimates its channel coefficient and the time that the BS receives the encoded channel coefficient transmitted by the user through uplink channel. However, compared to the feedback delay, if the channel variation is fast, there would be no need for the users to feed back the channel coefficients to the BS because the channels are different from what the BS already has. In this case, the power and rate allocation for downlink NOMA can be dependent on statistical CSI.

When the BS does not know the instantaneous CSI, it cannot control the transmit powers to meet the threshold SINRs as in (3.37). In addition, the code rates cannot be decided to adapt to instantaneous CSI. Accordingly, the transmit powers and rates need to be decided in advance and fixed regardless of instantaneous CSI.

We first derive the average (successful) transmission rate η_k or throughput of user k and then construct an optimization problem to control transmit power subject to a constraint of η_k. To begin with, let R_k denote the fixed transmission rate of user k when P_1 and P_2 are given. In addition, let $C_{l;k}$ be the instantaneous capacity for the signal to user l when decoded at user k. It is given by

$$C_{l;k} = \log_2\left(1 + \frac{\pi_k P_l}{\pi_k \sum_{m=1}^{l-1} P_m + 1}\right),$$ (3.56)

where $\pi_k = \frac{|h_k|^2}{N_0}$. For example, $C_{2;1}$ is the instantaneous capacity for the signal to user 2 at user 1. Thus, since user 1 must first decode the signal to user 2 for successful SIC, $C_{2;1}$ has to be greater than R_2.

At user 2, the probability of successful decoding is given by

$$\mathbb{P}_2 = \Pr(C_{2;2} > R_2)$$

$$= \Pr\left(\pi_2 > \frac{\tau_2}{P_2 - \tau_2 P_1}\right),$$ (3.57)

where $\tau_k = 2^{R_k} - 1$, if $P_2 > \tau_2 P_1$. Note that if $P_2 \le \tau_2 P_1$, $\mathbb{P}_2 = 0$. At user 1, the probability of successful decoding is slightly different from (3.57) as the signal to user 2 has to be decoded for SIC as well, which is given by

$$\mathbb{P}_1 = \Pr(C_{2;1} > R_2, C_{1;1} > R_1)$$

$$= \Pr\left(\pi_1 > \max\left\{\frac{\tau_2}{P_2 - \tau_2 P_1}, \frac{\tau_1}{P_1}\right\}\right).$$ (3.58)

Using the probability of successful decoding, the average transmission rate or throughput of each user can be found as

$$\eta_k = R_k \mathbb{P}_k.$$ (3.59)

It is interesting to note that if R_k is too high, the probability of successful decoding can be low, which results in a low throughput. On the other hand, if R_k is too low, although a high probability of successful decoding can be achieved, the throughput cannot be high due to a low transmission rate, R_k. Thus, the determination of R_k (together with the transmit power, P_k) plays a crucial role in the throughput maximization.

We assume that π_k is an independent exponential random variable (i.e. independent Rayleigh fading channels are assumed) and $\mathbb{E}[\pi_k] = \bar{\pi}_k$. Using (3.57), the throughput of user 2 becomes

$$\eta_2 = R_2 \Pr\left(\pi_2 > \frac{\tau_2}{P_2 - \tau_2 P_1}\right)$$

$$= R_2 \exp\left(-\frac{\tau_2}{\bar{\pi}_2(P_2 - \tau_2 P_1)}\right)$$

$$= \log_2(1 + \tau_2) \exp\left(-\frac{\tau_2}{\bar{\pi}_2(P_2 - \tau_2 P_1)}\right). \tag{3.60}$$

We can maximize η_2 with respect to τ_2 or R_2 for fixed P_1 and P_2.

On the other hand, the throughput of user 1 from (3.58) is given by

$$\eta_1 = \log_2(1 + \tau_1) e^{-\frac{1}{\bar{\pi}_1} \max\left(\frac{\tau_2}{P_2 - \tau_2 P_1}, \frac{\tau_1}{P_1}\right)}. \tag{3.61}$$

Unfortunately, the throughput maximization in this case is different from that for user 2 as η_1 depends on both τ_1 and τ_2. This demonstrates that the rate allocation with statistical CSI is not straightforward even if $\mathbf{p} = [P_1\ P_2]^T$ is given.

Since we assume that statistical CSI is available, the following problem can be formulated to minimize the total transmit power with individual throughput constraints:

$$\min_{P_k, R_k} \|\mathbf{p}\|_1$$

$$\text{subject to } \eta_k \geq \bar{\eta}_k,\ k = 1,2, \tag{3.62}$$

where $\|\mathbf{x}\|_1 = \sum_i |x_i|$ denotes the ℓ_1-norm of vector $\mathbf{x}$ and $\bar{\eta}_k$ is the target throughout of user k. From (3.59), η_k in (3.62) is expressed as

$$\eta_1 = R_1 \Pr(C_{2;1} > R_2, C_{1;1} > R_1) = R_1 \Pr(\pi_1 > \max(\omega_1, \omega_2)),$$
$$\eta_2 = R_2 \Pr(C_{2;2} > R_2) = R_2 \Pr(\pi_2 > \omega_2), \tag{3.63}$$

where $\omega_k = \frac{\tau_k}{P_k - \tau_k \sum_{l=1}^{k-1} P_l}$ for $k = \{1,2\}$. Note that (3.62) can be extended for $K > 2$.

The problem in (3.62) is involved as the rates should also be optimized. In other words, a joint rate and power allocation has to be considered for (3.62). It is important to note that since the throughput in (3.63) is not a convex function, the problem in (3.62) is not a convex problem. As a result, it seems difficult to find the solution as standard optimization solvers cannot be used. Fortunately, in Choi (2017a), a method to find the optimal solution is derived without using multidimensional optimization solvers for any $K \geq 2$.

Using the approach in Choi (2017a), the BS is able to jointly allocate rate and power to minimize the total transmit power with throughput constraints. For comparisons, we can also consider joint rate and power allocation for OMA. In OMA, the throughput of user k is given by

$$
\begin{aligned}
\eta_k &= \frac{1}{K} R_k \Pr\left(\log_2(1 + \alpha_k P_k) \geq R_k\right) \\
&= \frac{1}{K} \log_2(1 + \tau_k) e^{-\frac{\tau_k}{\alpha_k P_k}},
\end{aligned}
\tag{3.64}
$$

and the following optimization problem can be formulated:

$$
\min_{P_k, R_k} P_k
$$
$$
\text{subject to } \eta_k \geq \bar{\eta}_k.
\tag{3.65}
$$

To find the solution, consider the following sub-problem:

$$
\eta_k^*(P_k) = \max_{\tau_k \geq 0} \eta_k,
\tag{3.66}
$$

which is the maximum throughput for given power P_k. For convenience, denote by $R_k^*(P_k)$ the solution of R_k (via τ_k) of (3.66) that is a function of P_k. Then, the minimum power by joint rate and power allocation is given by

$$
P_k^* = \min\{P_k \mid \eta_k^*(P_k) \geq \bar{\eta}_k\}.
\tag{3.67}
$$

Note that once P_k^* is found, the optimal rate of (3.65) becomes $R_k^*(P_k^*)$.

Figure 3.9 shows the minimum total transmission powers of OMA and NOMA for a given set of target throughput values with $\{\bar{\pi}_1, \bar{\pi}_2\} = \{1, 1/4\}$. For any pair of target throughput values, $\{\bar{\eta}_1, \bar{\eta}_2\}$, we can confirm that the minimum total transmission power of NOMA is lower than that of OMA. In Figure 3.10a,b, we fix one target throughput and set various values for the other target throughput to see how the minimum total transmission power varies. As any target throughput increases, the minimum total transmission power increases. By comparing Figure 3.10a,b, we can observe that the increase of $\bar{\eta}_2$ (with fixed $\bar{\eta}_1 = 1$) requires a more increase of total transmission power than the increase of $\bar{\eta}_1$ (with fixed $\bar{\eta}_2 = 1$), which results from the fact that $\bar{\pi}_2$ is lower than $\bar{\pi}_1$.

Figure 3.11 shows the minimum total transmission powers of OMA and NOMA with target throughput values $\bar{\eta}_k = 1, k = 1, 2$, for different values of $\bar{\pi}_2$ when $\bar{\pi}_1 = 1$ is fixed. As the channel power gain of user 2, $\bar{\pi}_2$, increases, the total transmission power decreases in both OMA and NOMA. We can also see that the total transmission power of NOMA is always lower than that of OMA.

3.4 Code Division Multiple Access

In wireless communications, the channel resource is often given in the time and frequency domains. As a result, most OMA schemes are based on partitioning

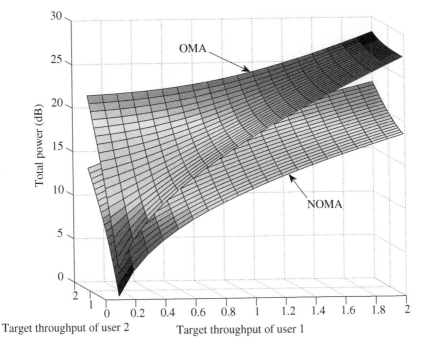

Figure 3.9 Surfaces of the minimum total transmission power, P_T, of OMA (the upper one) and NOMA (the lower one) for various values of target throughput, $(\bar{\eta}_1, \bar{\eta}_2) \in \{(x,y)|0 < x, y < 2\}$, when $\{\bar{\pi}_1, \bar{\pi}_2\} = \{1, 1/4\}$.

channel resources in the time or frequency domain (e.g. TDMA or FDMA, respectively). On the other hand, in (power-domain) NOMA, without partitioning channel resources, multiple users' signals can coexist in the same time and frequency domain, while a receiver differentiates them with their different power levels. In this sense, CDMA can be seen as another NOMA scheme, as the signals from multiple users coexisting in the same time and frequency domain can be differentiated by their different codes or signatures. In this section, we discuss CDMA and multiuser detection (Verdu, 1998; Choi, 2010).

3.4.1 DS-CDMA

In this section, we discuss a specific CDMA scheme, namely direct sequence code division multiple access (DS-CDMA), where multiple users' signals can be differentiated by their signature sequences.

In DS-CDMA systems, multiple users can share a common frequency band at the same time by using different signature waveform. In particular, each user

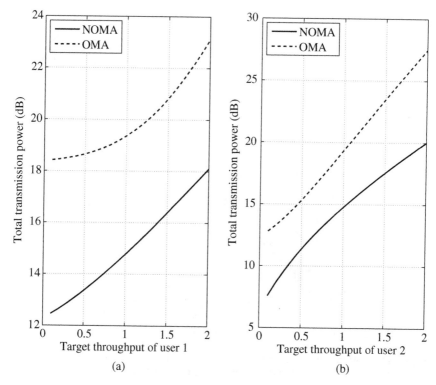

Figure 3.10 Minimum total transmission powers, P_T, of OMA and NOMA: (a) P_T for $\bar{\eta}_2 = 1$ and $\bar{\eta}_1 \in \{0,2\}$; (b) P_T for $\bar{\eta}_1 = 1$ and $\bar{\eta}_2 \in \{0,2\}$.

transmits its signal spread by the signature waveform, which is given by

$$s_k(t) = \sum_{m=0}^{N_{pg}-1} c_{m,k} p(t - mT_c), \ 0 \le t \le T, \tag{3.68}$$

where $s_k(t)$ and $\{c_{1,k}, c_{2,k}, \ldots, c_{N_{pg},k}\}$ are the signature waveform and the spreading sequence (or code) of user k, $p(t)$ is the chip waveform, and T_c is the chip interval. Here, T represents the symbol duration. In addition, we assume that $c_{m,k} \in \left\{ -\frac{1}{\sqrt{N_{pg}}}, \frac{1}{\sqrt{N_{pg}}} \right\}$ so that $\sum_{m=1}^{N_{pg}} |c_{m,k}|^2 = 1$ for the normalization purpose. It is generally assumed that $T = N_{pg} T_c$, where N_{pg} is called the processing gain. That is, there are N_{pg} chips per symbol interval. In addition, we assume that $p(t)$ satisfies

$$\int_{-\infty}^{\infty} p(t - nT_c)p(t - mT_c)dt = E_c \delta_{n,m}, \tag{3.69}$$

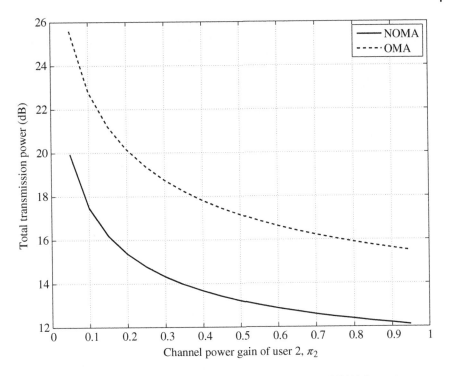

Figure 3.11 Minimum total transmission powers, P_T, of OMA and NOMA for various values of $\bar{\pi}_2$ when $\bar{\pi}_1 = 1$ with $\bar{\eta}_k = 1$, $k = 1,2$.

where E_c is the energy of $p(t)$. Here, $\delta_{n,m}$ is the Kronecker delta, which is defined as

$$\delta_{n,m} = \begin{cases} 1, & \text{if } n = m, \\ 0, & \text{otherwise} \end{cases}$$

Then, the received signal from K users over the AWGN channel in DS-CDMA is given by

$$r(t) = \sum_{k=1}^{K} A_k s_k(t) s_k + n(t), \tag{3.70}$$

where A_k is the amplitude of the kth user signal, and $n(t)$ is the background nose. Here, s_k represents the data symbol of user k. For convenience, suppose that $p(t)$ is a rectangular pulse with pulse width $T_\mathrm{c} = \frac{T}{N_\mathrm{pg}}$. The received signal is sampled at a rate of $\frac{1}{T_\mathrm{c}}$ after the filtering with $p(t)$. Then, with $y(t) = \frac{1}{\sqrt{E_\mathrm{c}}} \int p(\tau) r(t - \tau) d\tau$, we have

$$\mathbf{y} = [y(t_1) \cdots y(t_{N_\mathrm{pg}})]^\mathrm{T}$$

$$= \sum_{k=1}^{K} A_k \mathbf{c}_k s_k + \mathbf{n}, \tag{3.71}$$

where $\mathbf{c}_k = [c_{1,k} \cdots c_{N_{pg},k}]^T$, $\mathbf{n} = [n_1 \cdots n_{N_{pg}}]$ is the sampled noise vector, and $t_m = mT_c$, $m = 1, \ldots, N_{pg}$. Here $n_m = \frac{1}{\sqrt{E_c}} \int_{-\infty}^{\infty} p(mT_c - \tau)n(\tau)d\tau$. It can be further shown that

$$\mathbf{y} = \mathbf{CAs} + \mathbf{n}, \tag{3.72}$$

where $\mathbf{s} = [s_1 \, s_2 \cdots s_K]^T$, $\mathbf{n} = [n_1 \, n_2 \cdots n_{N_{pg}}]^T$, $\mathbf{A} = \mathrm{diag}(A_1, A_2, \ldots, A_K)$, and

$$\mathbf{C} = \begin{bmatrix} c_{1,1} & c_{1,2} & \cdots & c_{1,K} \\ c_{2,1} & c_{2,2} & \cdots & c_{2,K} \\ \vdots & \vdots & \ddots & \vdots \\ c_{N_{pg},1} & c_{N_{pg},2} & \cdots & c_{N_{pg},K} \end{bmatrix}.$$

The column vectors of $\mathbf{C}$ are the spreading codes of users. That is, $\mathbf{C} = [\mathbf{c}_1 \, \mathbf{c}_2 \cdots \mathbf{c}_K]$. Note that

$$\mathbb{E}[n_m n_q] = \frac{1}{E_c} \int_{-\infty}^{\infty} \int_{-\infty}^{\infty} \mathbb{E}[n(t)n(\tau)]p(t - mT_c)p(\tau - qT_c)dt \, d\tau$$
$$= \sigma^2 \delta_{m,q},$$

where $\sigma^2 = \frac{N_0}{2E_c}$. Hence, $\mathbf{n}$ is a Gaussian random noise vector of mean zero and covariance matrix $\mathbb{E}[\mathbf{nn}^T] = \sigma^2 \mathbf{I}$.

As shown in (3.72), the receiver is able to differentiate users using their different spreading or signature sequences. It is also interesting to see that DS-CDMA can be seen as a generalization of TDMA. To see this, assume that $N_{pg} = K$ and the $\mathbf{c}_k$'s are orthogonal, e.g. (scaled) standard basis vectors of N_{pg}-dimensional space. In this case, $\mathbf{C} = \frac{1}{\sqrt{N_{pg}}}\mathbf{I}$, which implies that one symbol interval is divided into N sub-symbol intervals, and the kth user can only transmit the kth sub-symbol interval. This becomes a TDMA system that supports $N_{pg} = K$ users.

In TDMA, if there is a user not transmitting a signal, that sub-symbol interval is wasted. To avoid this problem in DS-CDMA, spreading codes are not orthogonal. As a result, there is multi-user interference (MUI). To see this clearly, we can consider binary random sequences for spreading sequences, i.e. $\mathrm{Pr}\left(c_{n,k} = \frac{1}{\sqrt{N_{pg}}}\right) = \mathrm{Pr}\left(c_{n,k} = -\frac{1}{\sqrt{N_{pg}}}\right) = \frac{1}{2}$. In this case, $\mathbf{c}_k$ is seen as a random vector. At the receiver, if the signal from user k is to be detected, the correlator can be used as follows:

$$z_k = \mathbf{c}_k^T \mathbf{y}$$
$$= A_k s_k + \underbrace{\sum_{q \neq k} A_q \mathbf{c}_k^T \mathbf{c}_q s_q + \mathbf{c}_k^T \mathbf{n}}_{=\mathrm{MUI}}, \tag{3.73}$$

where z_k is the output of the correlator with weighting vector $\mathbf{c}_k$.

Example 3.2 The correlator is a key element of digital communications. There are different use cases. Suppose that there is only one transmitter that wants to send multiple bits with different signature waveforms. For example, consider four waveforms: $\psi_n(t) = \sum_{m=0}^{N_{pg}-1} c_{m,n} p(t - mT_c)$, $n = 0, \ldots, 3$. Then, it is possible to transmit two bits per T seconds. At the receiver, it is expected to receive one of four waveforms. For example, when the nth waveform is transmitted, a receiver wishes to detect the signal. To this end, the receiver can combine the received signal with weighting function, $w(t) = \sum_{m=0}^{N_{pg}-1} w_m p(t - mT_c)$, as follows:

$$
\begin{aligned}
z &= \int_0^T w(t)(\psi_n(t) + n(t))dt \\
&= \underbrace{E_c \sum_{m=0}^{N_{pg}-1} w_m c_{m,n}}_{=\text{signal}} + \underbrace{\int_0^T c(t)n(t)dt}_{=\text{noise}} \\
&= E_c \mathbf{w}^T \mathbf{c}_m + n,
\end{aligned}
$$

where $\mathbf{w} = [w_0 \cdots w_{N_{pg}-1}]^T$ and $n = \int_0^T c(t)n(t)dt \sim \mathcal{N}(0, E_c||\mathbf{w}||^2\sigma^2)$. Then, the SNR of the output of the correlator, z, becomes

$$
\text{SNR} = \frac{(E_c \mathbf{w}^T \mathbf{c}_m)^2}{E_c||\mathbf{w}||^2\sigma^2} \le \frac{E_c}{||\mathbf{w}||^2\sigma^2}||\mathbf{w}||^2||\mathbf{c}_m||^2 = \frac{E_c N_{pg}}{\sigma^2},
$$

where the inequality is due to the Cauchy–Schwarz inequality and the equality is achieved if $\mathbf{w} \propto \mathbf{c}_m$. That is, the SNR is maximized if $\mathbf{w} = \mathbf{c}_m$, and the resulting receiver is called the matched filter receiver or correlator receiver. CDMA can be seen as a generalization of this example.

For simplicity, let $s_k \in \{-1, +1\}$ (i.e. binary signaling) in (3.73). Then, for given $\mathbf{c}_k$, it can be shown that $\mathbb{E}[\mathbf{c}_k^T \mathbf{c}_q \mid \mathbf{c}_k] = 0$, and

$$
\mathbb{E}\left[\mathbf{c}_k^T \mathbf{c}_q \mathbf{c}_q^T \mathbf{c}_k \mid \mathbf{c}_k\right] = \mathbf{c}_k^T \left(\frac{1}{N_{pg}}\mathbf{I}\right)\mathbf{c}_k = \frac{1}{N_{pg}},
$$

$$
\mathbb{E}\left[\mathbf{c}_k^T \mathbf{c}_q \mathbf{c}_{q'}^T \mathbf{c}_k \mid \mathbf{c}_k\right] = 0, \quad q \ne q'. \tag{3.74}
$$

Thus, the mean of MUI is zero and the variance becomes $\frac{1}{N_{pg}}\sum_{q\ne k} A_q^2$. As a result, when the receiver is to detect the signal from user k, the SINR becomes

$$
\text{SINR}_k = \frac{A_k^2}{\frac{1}{N_{pg}}\sum_{q\ne k} A_q^2 + \sigma^2}, \tag{3.75}
$$

or if $A_k = A$ for all k, the SINR becomes

$$\text{SINR} = \frac{A^2}{\frac{K-1}{N_{\text{pg}}}A^2 + \sigma^2}$$

$$= \frac{N_{\text{pg}}}{K - 1 + \frac{\sigma^2}{A^2}} \approx \frac{N_{\text{pg}}}{K - 1}, \tag{3.76}$$

which shows that the SINR depends on the number of the users who share the same radio resource. This shows that, unlike TDMA, if some users do not transmit, the SINR is improved for other users transmitting the signal, so shared resources are not wasted.

3.4.2 Multiuser Detection Approaches

The approach based on the correlator in (3.73) can be used to detect a signal from a specific user, i.e. user k. This approach is regarded as an optimal one if the MUI vector, $\sum_{q \neq k} A_q \mathbf{c}_q s_q$, is white, i.e. its covariance matrix is a scaled identity matrix. In fact, if the elements of $\mathbf{c}_q$ are considered to be iid, the MUI vector is white and the correlator maximizes the SINR. However, in practice, each user has a certain spreading or signature sequence that is known to the receiver. In other words, $\mathbf{C}$ is known to the receiver and the approach based on the correlator in (3.73) is no longer optimal.

With known $\mathbf{C}$, there are various optimal detectors that allow to detect all K signals simultaneously, which are called multiuser detectors.

An optimal detector based on the maximum likelihood (ML) principle can be considered. With a known amplitude matrix, $\mathbf{A}$, and a spreading code matrix, $\mathbf{C}$, the likelihood function is given by

$$f(\mathbf{y}|\mathbf{s}) = C \exp\left(-\frac{1}{2\sigma^2}||\mathbf{y} - \mathbf{CAs}||^2\right), \tag{3.77}$$

where C is a normalizing constant. Let $s_k \in S$, where S is the signal constellation. Then, the ML solution becomes

$$\mathbf{s}_{\text{ml}} = \arg\max_{\mathbf{s} \in S^K} f(\mathbf{y}|\mathbf{s})$$

$$= \arg\min_{\mathbf{s} \in S^K} ||\mathbf{y} - \mathbf{CAs}||^2, \tag{3.78}$$

where S^K is the K-fold Cartesian product of S.

The complexity to find the ML solution depends on the size of the search space, i.e. $|S^K| = |S|^K$. Thus, the complexity grows exponentially with K. Due to a high complexity, the ML detector may be impractical. Therefore, suboptimal but less complex approaches can be considered. Among those, the decorrelating detector is a suboptimal detector to suppress MUI using a linear transform.

As in (3.73), consider the outputs of the correlators:

$$z_k = \mathbf{c}_k^{\mathrm{T}} \mathbf{y}, \ k = 1, 2, \dots, K. \tag{3.79}$$

We have the desired user's output as well as the interfering signals from the other users. Stacking the z_k's, we have

$$
\begin{aligned}
\mathbf{z} &= [z_1 \ z_2 \ \cdots \ z_K]^{\mathrm{T}} \\
&= \mathbf{C}^{\mathrm{T}} \mathbf{y} \\
&= \mathbf{C}^{\mathrm{T}} \mathbf{C} \mathbf{A} \mathbf{s} + \mathbf{C}^{\mathrm{T}} \mathbf{n}. \tag{3.80}
\end{aligned}
$$

To suppress the MUI, a linear transform

$$\mathbf{R}^{-1} = \left(\mathbf{C}^{\mathrm{T}} \mathbf{C}\right)^{-1}$$

results in a zero-MUI such as

$$
\begin{aligned}
\mathbf{d} &= \mathbf{R}^{-1} \mathbf{z} \\
&= \mathbf{A} \mathbf{s} + \mathbf{R}^{-1} \mathbf{C}^{\mathrm{T}} \mathbf{n}. \tag{3.81}
\end{aligned}
$$

If the desired user is user k, the kth element of $\mathbf{d}$ becomes the signal of interest:

$$d_k = [\mathbf{d}]_k = A_k s_k + [\mathbf{R}^{-1} \mathbf{C}^{\mathrm{T}} \mathbf{n}]_k. \tag{3.82}$$

Even though no MUI exists, the decorrelating detector can have a limited performance if the spreading codes are highly correlated. To see the impact of spreading codes' cross-correlation, consider the SNR of the signal in (3.82),

$$\mathrm{SNR}_{\mathrm{dc},k} = \frac{A_k^2}{\sigma^2 [\mathbf{R}^{-1}]_{k,k}} \tag{3.83}$$

since

$$
\begin{aligned}
\mathbb{E}[([\mathbf{R}^{-1} \mathbf{C}^{\mathrm{T}} \mathbf{n}]_k)^2] &= \left[\mathbb{E}\left[\left(\mathbf{R}^{-1} \mathbf{C}^{\mathrm{T}} \mathbf{n}\right)\left(\mathbf{R}^{-1} \mathbf{C}^{\mathrm{T}} \mathbf{n}\right)^{\mathrm{T}} \right] \right]_{k,k} \\
&= \left[\mathbf{R}^{-1} \mathbf{C}^{\mathrm{T}} E[\mathbf{n}\mathbf{n}^{\mathrm{T}}] \mathbf{C} \mathbf{R}^{-1} \right]_{k,k} \\
&= \sigma^2 [\mathbf{R}^{-1}]_{k,k}. \tag{3.84}
\end{aligned}
$$

If there are highly correlated spreading codes, the diagonal elements of $\mathbf{R}^{-1}$ become large, which results in a low SNR.

From (3.72), the linear minimum mean squared error (LMMSE) detector can be found. We assume that the detector is linear and is to estimate the signal vector $\mathbf{A}\mathbf{s}$. Then, the LMMSE detector is given by

$$
\begin{aligned}
\mathbf{L}_{\mathrm{lm}} &= \arg \min_{\mathbf{L}} \mathbb{E}[\|\mathbf{A}\mathbf{s} - \mathbf{L}\mathbf{y}\|^2] \\
&= \mathbf{R}_{\mathbf{A}\mathbf{s},\mathbf{y}} \mathbf{R}_{\mathbf{y}}^{-1}, \tag{3.85}
\end{aligned}
$$

where $\mathbf{R}_y = \mathbb{E}[\mathbf{y}\mathbf{y}^T]$ is the covariance matrix of $\mathbf{y}$ and $\mathbf{R}_{\mathbf{As},y} = \mathbb{E}[\mathbf{As}\mathbf{y}^T]$. Assuming that the elements of $\mathbf{s}$ are iid binary random variables and independent of $\mathbf{n}$, we can show that

$$\mathbf{R}_{\mathbf{As},y} = \mathbf{A}^2\mathbf{C}^T,$$
$$\mathbf{R}_y = \mathbf{C}\mathbf{A}^2\mathbf{C}^T + \sigma^2\mathbf{I}.$$

Thus, the LMMSE detector is written as

$$\mathbf{L}_{lm} = \mathbf{A}^2\mathbf{C}^T(\mathbf{C}\mathbf{A}^2\mathbf{C}^T + \sigma^2\mathbf{I})^{-1}. \tag{3.86}$$

The correlation matrix of the error, $\mathbf{As} - \mathbf{L}_{lm}\mathbf{y}$, is written as

$$\begin{aligned} \mathbf{R}_{lm} &= \mathbb{E}[(\mathbf{As} - \mathbf{L}_{lm}\mathbf{y})(\mathbf{As} - \mathbf{L}_{lm}\mathbf{y})^T] \\ &= \mathbf{A}^2 - \mathbf{A}^2\mathbf{C}^T(\mathbf{C}\mathbf{A}^2\mathbf{C}^T + \sigma^2\mathbf{I})^{-1}\mathbf{C}\mathbf{A}^2 \\ &= \mathbf{A}^2\left(\mathbf{I} - \mathbf{A}^2\mathbf{C}^T(\mathbf{C}\mathbf{A}^2\mathbf{C}^T + \sigma^2\mathbf{I})^{-1}\mathbf{C}\right) \\ &= \left[\mathbf{A}^{-2} + (\sigma^2)^{-1}\mathbf{C}^T\mathbf{C}\right]^{-1}, \end{aligned} \tag{3.87}$$

where the last equality is obtained by using the matrix inversion lemma. If $\mathbf{A} = \mathbf{I}$, $\mathbf{R}_{lm}$ is simplified to

$$\begin{aligned} \mathbf{R}_{lm} &= \mathbf{I} - \mathbf{C}^T(\mathbf{C}\mathbf{C}^T + \sigma^2\mathbf{I})^{-1}\mathbf{C} \\ &= \left[\mathbf{I} + (\sigma^2)^{-1}\mathbf{C}^T\mathbf{C}\right]^{-1}. \end{aligned} \tag{3.88}$$

Note that the LMMSE detector can be considered as $\mathbf{s}$ being the desired signal vector rather than $\mathbf{As}$. With $\mathbf{s}$ as the desired signal, the LMMSE detector becomes

$$\overline{\mathbf{L}}_{lm} = \mathbf{A}\mathbf{C}^T(\mathbf{C}\mathbf{A}^2\mathbf{C}^T + \sigma^2\mathbf{I})^{-1} = \mathbf{A}^{-1}\mathbf{L}_{lm} \tag{3.89}$$

and the correlation matrix of the error vector, $\mathbf{s} - \mathbf{L}_{lm}\mathbf{y}$, is given by

$$\begin{aligned} \overline{\mathbf{R}}_{lm} &= \left[\mathbf{I} + (\sigma^2)^{-1}\mathbf{A}\mathbf{C}^T\mathbf{C}\mathbf{A}\right]^{-1} \\ &= \mathbf{A}^{-1}\mathbf{R}_{lm}\mathbf{A}^{-1}. \end{aligned} \tag{3.90}$$

There is no difference between $\mathbf{L}_{lm}$ and $\overline{\mathbf{L}}_{lm}$ except scaling by the amplitude matrix, $\mathbf{A}$.

From (3.89), the LMMSE estimate of s_k becomes

$$\begin{aligned} \hat{s}_{lm,k} &= \mathbf{w}_k^T\mathbf{y} \\ &= A_k\mathbf{c}_k^T\mathbf{R}_y^{-1}\mathbf{y} \\ &= A_k^2\mathbf{c}_k^T\mathbf{R}_y^{-1}\mathbf{c}_k s_k + A_k\mathbf{c}_k^T\mathbf{R}_y^{-1}\left(\sum_{q\neq k}A_q\mathbf{c}_q s_q + \mathbf{n}\right), \end{aligned} \tag{3.91}$$

where $\mathbf{w}_k = \mathbf{R}_\mathbf{y}^{-1}\mathbf{c}_k A_k$ is the LMMSE weighting vector for user k. From (3.90), the mean squared error (MSE) of $\hat{s}_{\mathrm{lm},k}$ is written as

$$\mathrm{MSE}_k = 1 - A_k^2 \mathbf{c}_k^\mathrm{T} \mathbf{R}_\mathbf{y}^{-1} \mathbf{c}_k. \tag{3.92}$$

Then, the SINR of $s_{\mathrm{lm},k}$ can be found in terms of MSE_k as follows:

$$
\begin{aligned}
\mathrm{SINR}_k &= \frac{|A_k^2 \mathbf{c}_k^\mathrm{T} \mathbf{R}_\mathbf{y}^{-1} \mathbf{c}_k|^2}{A_k^2 \mathbf{c}_k^\mathrm{T} \mathbf{R}_\mathbf{y}^{-1} (\mathbf{R}_\mathbf{y} - A_k^2 \mathbf{c}_k \mathbf{c}_k^\mathrm{T}) \mathbf{R}_\mathbf{y}^{-1} \mathbf{c}_k} \\
&= \frac{A_k^2 \mathbf{c}_k^\mathrm{T} \mathbf{R}_\mathbf{y}^{-1} \mathbf{c}_k}{1 - A_k^2 \mathbf{c}_k^\mathrm{T} \mathbf{R}_\mathbf{y}^{-1} \mathbf{c}_k} \\
&= \frac{1 - \mathrm{MSE}_k}{\mathrm{MSE}_k}. \tag{3.93}
\end{aligned}
$$

This justifies the LMMSE detector as it is to minimize MSE, which leads to maximize the SINR as shown in (3.93). Under Gaussian approximations, the achievable rate for each user can be found as $R_k = \log_2(1 + \mathrm{SINR}_k) = \log_2 \left(\frac{1-\mathrm{MSE}_k}{\mathrm{MSE}_k} \right)$.

In DS-CDMA, although MUD is employed, it is required that the number of users should be less than or equal to the processing gain, N_{pg}, i.e. $K \le N_{\mathrm{pg}}$, for a reasonably high SINR. In addition, it is expected that each user has a unique signature sequence. Consequently, the number of users, K, is limited to the processing gain, N_{pg}. Recall that we assumed that $T_\mathrm{c} = \frac{T}{N_{\mathrm{pg}}}$, where T is the symbol duration. For a fixed symbol duration, T, the increase of N_{pg} implies the decrease of T_c, which leads to the increase of the system bandwidth. To see this clearly, consider (3.68). The bandwidth of the signature waveform is proportional to that of $p(t)$. If $p(t)$ is a rectangular pulse of width T_c, we can see that the bandwidth is proportional to $\frac{1}{T_\mathrm{c}}$, which means that the bandwidth increases with N_{pg} for a fixed T. In other words, in order to increase the number of users, K, the system bandwidth has to increase, like TDMA.

Interestingly, the above observations do not strongly advocate the advantages of DS-CDMA over any OMA scheme (e.g. TDMA), i.e. so shared resources are not wasted if some users do not transmit in DS-CDMA. This problem arises as we assumed that the number of users, K, is fixed, and that all users always transmit. As will be shown in Section 4.6, if DS-CDMA is used for random access, where the number of active users is varying, the advantages of DS-CDMA can be well exploited.

3.5 Further Reading

More information on the various multiple access schemes used for wireless communication can be found in Tse and Viswanath (2005). Since power-domain

NOMA is rooted in information theory, it is crucial to understand fundamentals of information theory. The reader is referred to Cover and Thomas (2006) or other textbooks on information theory. There are reviews on NOMA in Dai et al. (2015), Ding et al. (2017), and Dai et al. (2018). For the multiuser detection in CDMA, the reader is referred to Verdu (1998).

4

Random Access Systems

In the previous chapter, we considered OMA schemes such as time division multiple access (TDMA), frequency division multiple access (FDMA), and orthogonal frequency division multiple access (OFDMA), in which a fixed number of users share a given channel resource based on a specific channel allocation mechanism. In general, it was assumed that each user continuously transmits signals through a dedicated (orthogonal) channel for uplink transmissions, and the channel allocation can be carried out based on user's requirements (e.g. the data rate) in advance. The resulting transmissions by multiple users through their dedicated channels are referred to as coordinated transmissions. Certainly, the signaling overhead to allocate dedicated channels to multiple users in coordinated transmissions is negligible if users have continuous data streams to transmit.

However, if each user occasionally has a small amount of data to transmit or has a low duty cycle, the signaling overhead to allocate dedicated channels becomes overwhelming compared to actual amount of data, leading to a low spectral efficiency. To minimize the signaling overhead, uncoordinated transmissions are preferable, where a user with data can transmit through a common shared channel whenever necessary. However, collisions can happen if multiple users attempt to transmit simultaneously. Clearly, upon collision, the receiver fails to decode signals. Therefore, uncoordinated transmissions are desirable when the signaling overhead reduction offsets the performance degradation due to collisions.

In this chapter, we discuss well-known random access protocols that are used for uncoordinated transmissions by users of low duty cycle or sparse activity transmitting a small amount of data.

Massive Connectivity: Non-Orthogonal Multiple Access to High Performance Random Access,
First Edition. Jinho Choi.
© 2022 The Institute of Electrical and Electronics Engineers, Inc. Published 2022 by John Wiley & Sons, Inc.

4.1 ALOHA Systems

In this section, we mainly discuss ALOHA system that was developed by the University of Hawaii as a computer networking system in 1970s. ALOHA is known to stand for Additive Links On-line Hawaii Area.

4.1.1 Single Channel Random Access

Suppose that there are multiple users who do not transmit continuously. In particular, each user does not always have a packet to send, but very occasionally. As a result, multiple users can share a given channel, and uncoordinated transmissions can be used to keep signaling overhead low. For example, consider Figure 4.1 where four users transmit their packets to a receiver, which is a base station (BS), through a shared channel. Let t_k be the time for user k to start to transmit a data packet and δ_k its length. Accordingly, $t_k + \delta_k$ becomes the time that user k completes the packet transmission. Assuming that $t_k < t_{k+1}$, if $t_{k+1} > t_k + \delta_k$, the BS can receive all the packets from four users without any overlapping transmission (i.e. collision). As long as the packet length is sufficiently short and packets are well distributed over time, multiple users can share a single channel to transmit their packets successfully without coordinated transmission. However, due to the lack of coordination, collision (overlapping) can happen. As shown in Figure 4.1, since $t_3 < t_2 + \delta_2$ (i.e. user 3 starts its packet transmission before user 2 completes its transmission), the data packets from users 2 and 3 collide. The resulting approach is called pure ALOHA.

In pure ALOHA, users are not synchronized. Thus, collisions happen even if only a small portion of a packet is overlapped with another packet. To avoid this problem, suppose that users are synchronized. In addition, thanks to synchronization, each user with data can start transmitting only at the beginning of a time slot,

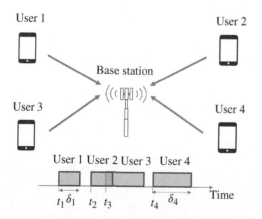

Figure 4.1 An illustration of uncoordinated transmissions by four users.

Figure 4.2 An illustration of
S-ALOHA with two users.

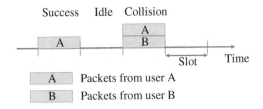

assuming that the packet size is equal to the length of the time slot. The resulting ALOHA system is referred to as slotted-Additive Links On-line Hawaii Area or S-ALOHA.

For S-ALOHA, consider a time-slotted system as illustrated in Figure 4.2, where time is divided into discrete slots. Users transmit their packets to a slot and any transmission attempt takes up exactly one time slot. A packet transmission of a user becomes successful if there are no other users transmitting packets to the same slot. As shown in Figure 4.2, each slot can have one of the following states: idle, success, and collision. Then, the state of S-ALOHA becomes idle, success, or collision if the number of transmitted packets is 0, 1, or k, respectively, where $k \geq 2$.

4.1.2 Multichannel S-ALOHA

In S-ALOHA, multiple users share a channel, referred to as a single-channel ALOHA. A generalization of single-channel ALOHA is multichannel ALOHA where multiple (independent) channels are shared by users. To this end, OMA schemes (e.g. TDMA or FDMA) can be used. As we shall see that as the number of users increases, it is necessary to increase the number of channels to maintain a reasonable performance. In terms of scalability, the notion of multichannel ALOHA is thus important, although it looks straightforward.

Suppose L channels in a multichannel ALOHA system. If a user has a data packet to send, the user can choose one of L channels uniformly at random. It can be seen as a group of L S-ALOHA systems where a user can randomly choose one of L S-ALOHA systems for each transmission of a data packet.

4.2 Throughput Analysis

In this section, we study the throughput of ALOHA protocols under the assumption that all collided packets are dropped. This assumption implies that it is not necessary for the active users that transmit packets to know whether or not packets are successively transmitted without collision (as collided packets will be ignored/dropped by the receiver). As a result, no feedback from the receiver is required.

Another important assumption is that the SNR is sufficiently high so that any transmitted packets can be successfully decoded if no collisions happen.

4.2.1 Pure ALOHA

In order to find the throughput of pure ALOHA, we assume as follows:

- The length of packet, denoted by T, is the same. Here, T can be seen as the unit time.
- The number of active users to send packets follows a Poisson distribution. The average number of active users during T is denoted by G.

Suppose that a user transmits a packet at time t. This packet can be successfully transmitted if there are no other packets within the time window of $(t - T, t + T)$. The average number of active users with the time window of $(t - T, t + T)$. Since the number of active users follows a Poisson distribution, the probability that there are no other active users is e^{-2G}, while the rate of transmission attempts by any active users within the unit time is G. Thus, the throughput that is the average number of successful packet transmissions per unit time becomes

$$\eta_{\text{pure}} = G\, e^{-2G}. \tag{4.1}$$

Note that since any collided packets are dropped, for the throughput, we only need to consider the packets that can be successfully transmitted without collision. The maximum can be obtained by taking the derivative with respect to G and setting it zero, because η_{pure} is a $\cap$-shape function in G. The throughput is maximized when $G = \frac{1}{2}$ and the maximum throughput becomes $\frac{1}{2}e^{-1}$.

4.2.2 Slotted ALOHA

In S-ALOHA, we assume that the length of slot is T, which is also equivalent to that of packets. Again, it is assumed that the number of active users, denoted by K, follows a Poisson distribution. The average number of active users per unit time (i.e. T) is G. Then, the probability of successful packet transmissions per unit time becomes the probability that there is only one active user, which is given by

$$\eta_{\text{slotted}} = \Pr(K = 1) = G\, e^{-G}. \tag{4.2}$$

In general, $f(x) = x\, e^{-x}$ is a well studied function: it is not difficult to show that the maximum is e^{-1}, which is achieved if $x = 1$. Thus, the maximum throughput of S-ALOHA is $e^{-1} \approx 0.3678$, which is two times higher than that of pure ALOHA.

To achieve the maximum throughput, it is expected that there is one active user per slot on average.

Example 4.1 The throughput in (4.2) is also an approximation when the total number of user is finite. To see this, suppose that there are M users and each user becomes active with a probability of q_a. Then, the number of active users, K, can follow the following binomial distribution:

$$K \sim \text{Bin}(M, q_a)$$

or $\Pr(K = k) = \binom{M}{k} q_a^k (1 - q_a)^{M-k}$. In this case, the throughput, which is the probability that only one user is active, is given by

$$\eta_{\text{slotted}}(M, q_a) = \Pr(K = 1) = M q_a (1 - q_a)^{M-1}. \tag{4.3}$$

The maximum throughput can be achieved when $q_a = \frac{1}{M}$, $M \geq 1$. Suppose that $M q_a$ is fixed when M grows. For a sufficiently large M, we have $(1 - q_a)^{M-1} \approx e^{-(M-1)q_a} \approx e^{-M q_a}$. Since the average number of active devices is $M q_a$, letting $G = M q_a$, we have

$$\eta_{\text{slotted}}(M, q_a) \to G e^{-G}, \quad M \to \infty,$$

which is identical to (4.2). In Figure 4.3, the (maximum) throughput in (4.3) is shown with $q_a = \frac{1}{M}$. It is noteworthy that the asymptotic maximum throughput is e^{-1} as $G = 1$.

4.2.3 Multichannel ALOHA

Suppose that there are L channels and each active user can choose one of them uniformly at random. Let K_l denote the number of active users choosing channel l. Recalling that K denotes the number of active users, we have

$$K = K_1 + \cdots + K_L.$$

It would be reasonable to assume that the K_l's are iid. Since K is assumed to be a Poisson random variable with mean G, we have that each K_l is a Poisson random variable with mean $\frac{G}{L}$ (note that the sum of Poisson random variables is also a Poisson random variable). Thus, the throughput of multichannel ALOHA becomes

$$\eta_{\text{MA}} = \sum_{l=1}^{L} \Pr(K_l = 1)$$

$$= \sum_{l=1}^{L} \frac{G}{L} e^{-\frac{G}{L}}$$

$$= G e^{-\frac{G}{L}}. \tag{4.4}$$

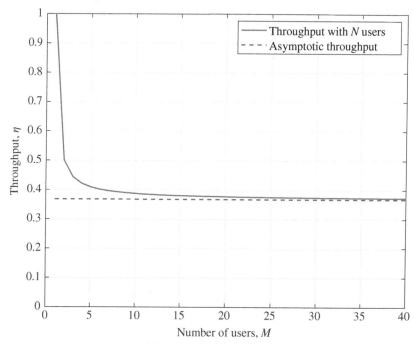

Figure 4.3 Maximum throughput of S-ALOHA as a function of M with $G = 1$ when the number of users, M, is finite.

Noting that the throughput can also be expressed as $\eta_{MA} = L x\, e^{-x}$, where $x = \frac{G}{L}$, we can see that the maximum throughput becomes $L\, e^{-1}$, i.e. L-time higher than that of S-ALOHA. This observation is not surprising as multichannel ALOHA has L-time more channel resource than single-channel S-ALOHA.

The throughput in (4.4) can also be found using a different way. Suppose that there are K active users and each active user can choose one of L channels uniformly at random. Since all the active users have the same conditions, we only need to consider one active user, say user 1. Assume that user 1 chooses channel 1. Then, user 1 does not have any collision if the other $K - 1$ active users choose different channels. Since each active user chooses a channel independently, the probability for this event to happen is

$$\underbrace{\left(1 - \frac{1}{L}\right) \times \cdots \times \left(1 - \frac{1}{L}\right)}_{(K-1)\ \text{times}} = \left(1 - \frac{1}{L}\right)^{K-1},$$

where $1 - \frac{1}{L}$ is the probability that another active user chooses one of channels $2, \ldots, L$. As a result, the average number of the active users without collisions is given by

$$\eta_{MA}(K) = K\left(1 - \frac{1}{L}\right)^{K-1},$$ (4.5)

which can be seen as the conditional throughput (for given K active users). Then, the throughput becomes $\eta_{MA} = \mathbb{E}[\eta_{MA}(K)]$. If K follows a Poisson distribution with mean G, it can be shown that

$$\begin{aligned}
\eta_{MA} &= \sum_{k=0}^{\infty} \eta_{MA}(k) \Pr(K = k) \\
&= \sum_{k=0}^{\infty} k\left(1 - \frac{1}{L}\right)^{k-1} \frac{e^{-G} G^k}{k!} \\
&= G e^{-G} \sum_{k=1}^{\infty} \frac{\left(\left(1 - \frac{1}{L}\right) G\right)^{k-1}}{(k-1)!} \\
&= G e^{-G} e^{G\left(1 - \frac{1}{L}\right)} \\
&= G e^{-\frac{G}{L}},
\end{aligned}$$ (4.6)

which is identical to (4.4). Note that the fourth equality in (4.6) is due to $e^x = \sum_{k=0}^{\infty} \frac{x^k}{k!}$.

If the number of users is finite, we may need to consider a binomial distribution for K. For example, suppose that there are a total of M users and each user becomes active with a probability of q_a. Then, it can be shown that

$$\begin{aligned}
\eta_{MA} &= \mathbb{E}[\eta_{MA}(K)] \\
&= \sum_{k=0}^{M} k\left(1 - \frac{1}{L}\right)^{k-1} \binom{M}{k} q_a^k (1 - q_a)^{M-k} \\
&= M q_a \sum_{k=1}^{M} \frac{(M-1)!}{(k-1)!(M-k)!} \left(q_a\left(1 - \frac{1}{L}\right)\right)^{k-1} (1 - q_a)^{M-1-(k-1)} \\
&= M q_a \left(1 - \frac{q_a}{L}\right)^{M-1}.
\end{aligned}$$ (4.7)

Since $1 - x \approx e^{-x}$ for $|x| \ll 1$, for a large L, the throughput becomes

$$\eta_{MA} \approx M q_a e^{-\frac{M q_a}{L}} = G e^{-\frac{G}{L}},$$ (4.8)

where $G = M q_a$ is the average number of active users per slot. This shows that even for a limited number of users, the same throughput can be achieved as for an infinite number of users.

In Figure 4.4, the throughput curves of pure, slotted, and multichannel ALOHA protocols are shown as functions of G. For multichannel ALOHA, we assume that $L = 2$. It is shown that the throughput of S-ALOHA is higher than that of pure

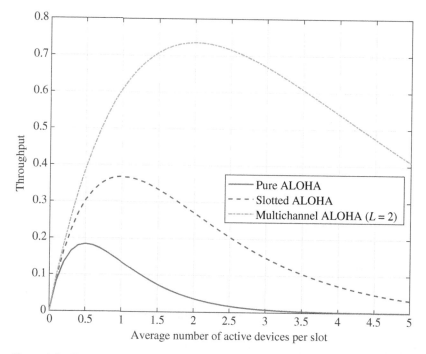

Figure 4.4 Throughput curves of three different ALOHA protocols as functions of *G*.

ALOHA thanks to synchronization. In addition, since multichannel ALOHA has two2 times more channel resource than S-ALOHA, it is clearly shown that it has a higher throughput than S-ALOHA.

However, if *L* multiple channels are formed by dividing one channel resource block into *L* sub-channel resource blocks, the maximum throughput of multichannel ALOHA becomes that of S-ALOHA. To see this, consider a session consisting of *L* time slots, and each slot is used as a channel for multichannel ALOHA, where an active user can choose one of *L* slots for random access. In this case, the average number of active users per session becomes *GL*. As a result, from (4.4), the throughput of multichannel ALOHA becomes

$$\eta_{\text{MA}} = \frac{(GL)e^{-\frac{GL}{L}}}{L} = G\,e^{-G}. \tag{4.9}$$

Here, the denominator becomes *L* as there are *L* slots per session. This clearly shows that the throughput of multichannel ALOHA is identical to that of S-ALOHA if the same size of resource block is used. This observation suggests that multichannel ALOHA may not be useful. However, there are some aspects that multichannel ALOHA can be useful. One of them is fast retrial, which will be discussed in Section 4.5.

4.3 Analysis with a Finite Number of Users

In Section 4.2, we simply assumed that collided packets are dropped to see the throughput. In this section, we consider the case that collided packets are to be retransmitted. In this case, there should be feedback from the receiver to active users and there are two different types of feedback, namely positive acknowledgment (ACK) for successful decoding and negative acknowledgment (NACK) for unsuccessful decoding. Each user needs to keep collided packets for retransmissions. Thus, the state of the system (of all users) can be represented by the number of backlogged packets (due to collisions).

4.3.1 A Markov Chain

Let M be the number of users in the system. At the beginning of slot t, suppose that there are N users with backlogged packets, where $N \leq M$. We assume that users with backlogged packets do not generate any packets to transmit. Thus, N is also the number of backlogged packets and $N \in \{0, \ldots, M\}$. It can be shown that the number of backlogged packets in slot $t + 1$ depends on that in slot t, which means that N is a Markov chain with a finite state space. The state transition probability of N is obtained in this subsection.

Let q_r and q_a denote the probability of retransmission (of users with backlogged packets) and the probability of transmission (of users with new packets), respectively. It should be noted that q_r cannot be 1. If $q_r = 1$, any users with collided packets will retransmit in the next slot. Since there are multiple users with collided packets when packet collision happens, immediate retransmissions result in subsequent packet collision. Thus, no users can succeed to transmit their packets. The state of the system is N and the probability that i users with new packets transmit is given by

$$P_a(i \mid N) = \binom{M - N}{i} q_a^i (1 - q_a)^{M-N-i}, \tag{4.10}$$

while the probability that i users with backlogged packets transmit is given by

$$P_r(i \mid N) = \binom{N}{i} q_r^i (1 - q_r)^{N-i}. \tag{4.11}$$

Then, the state transition probability, which is the probability that state N in slot t moves state $N + i$ in slot $t + 1$, is given by

$$P_{N,N+i} = \begin{cases} P_a(i \mid N), & \text{if } 2 \leq i \leq M - N, \\ P_a(1 \mid N)(1 - P_r(0 \mid N)), & \text{if } i = 1, \\ P_a(1 \mid N)P_r(0 \mid N) + P_a(0 \mid N)(1 - P_r(1 \mid N)), & \text{if } i = 0, \\ P_a(0 \mid N)P_r(1 \mid N), & \text{if } i = -1. \end{cases} \tag{4.12}$$

For $2 \le i \le M - N$ (note that i cannot be greater than $M - N$ as the current state is N and $N \in \{0, \ldots, M\}$ as mentioned earlier), i new packets are to be collided so that the system has $N + i$ backlogged packets. To this event, there should be i new packets regardless of retransmissions from backlogged users. For $i = 1$, one new packet is to be transmitted and collided with retransmitted backlogged packets. The corresponding probability is the product of $P_a(1 \mid N)$ and $1 - P_r(0 \mid N)$. It is also straightforward to find the state transition probability when $i = 0$ and $i = -1$ in (4.12). Note that i cannot be less than -1 as up to one packet per slot can be successfully transmitted in S-ALOHA.

As shown in (4.12), thanks to a finite number of users, M, in the system, we are able to find the state transition probability. From this, the stationary distribution of N can also be found. However, unfortunately, we cannot obtain much insight into the behavior of the system from this.

4.3.2 Drift Analysis

For a Markov chain, we can define the drift as the expected change over one slot time, starting in state $N = n \in \{0, \ldots, M\}$. Let N_t denote the state in slot t. Then, the drift is formally defined as

$$D_n = \mathbb{E}[N_{t+1} - N_t \mid N_t = n]$$
$$= \mathbb{E}[N_{t+1} \mid N_t = n] - n. \tag{4.13}$$

If $D_n > 0$ or $D_n < 0$, the number of backlogged packets increases or decreases, respectively, on average. Thus, for a stable system, we expect that there exists an $n^* > 0$ such that $D_n < 0$ for $n > n^*$ and $D_n > 0$ for $n < n^*$.

From (4.12), it can be shown that

$$D_n = \sum_{i=-1}^{M-n} i P_{n,n+i}$$
$$= \sum_{i=0}^{M-n} i P_a(i \mid n) - P_a(1 \mid n) P_r(0 \mid n) - P_a(0 \mid n) P_r(1 \mid n). \tag{4.14}$$

In (4.14), from (4.10), we have $\sum_{i=0}^{M-n} i P_a(i \mid n) = (M - n) q_a$. Thus, the drift in (4.14) can be rewritten as

$$D_n = (M - n) q_a - P_{\text{succ}}, \tag{4.15}$$

where P_{succ} is the probability that a packet can be successfully transmitted, which is given by

$$
\begin{aligned}
P_{\text{succ}} &= P_a(1 \mid n) P_r(0 \mid n) + P_a(0 \mid n) P_r(1 \mid n) \\
&= (M - n) q_a (1 - q_a)^{M-n-1} (1 - q_r)^n + n(1 - q_a)^{M-n} q_r (1 - q_r)^{n-1} \\
&\approx \left((M - n) q_a + n q_r \right) e^{-((M-n)q_a + n q_r)},
\end{aligned}
\tag{4.16}
$$

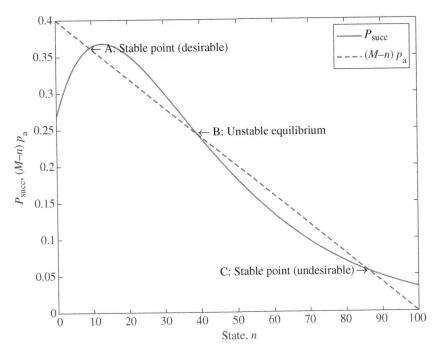

Figure 4.5 Two curves as functions of state n, $(M - n)q_a$ and $P_{succ} = G(n)e^{-G(n)}$, when $M = 100$, $q_a = 0.004$, and $q_r = 0.05$.

where the approximation is reasonable when $q_a, q_r \ll 1$ (using $1 - x \approx e^{-x}$ for $0 \le x \ll 1$). In state n, the average number of users transmitting is $G(n) = (M - n)q_a + nq_r$. Thus, we have $P_{succ} \approx G(n)e^{-G(n)}$, and

$$D_n \approx (M - n)q_a - G(n)e^{-G(n)}. \tag{4.17}$$

In Figure 4.5, the two functions of state n, $(M - n)q_a$ and $P_{succ} = G(n)e^{-G(n)}$, are shown when $M = 100$, $q_a = 0.004$, and $q_r = 0.05$. In general, q_r is larger than q_a to avoid a long access delay. The difference between the two curves in Eq. (4.15) is the drift as shown in (4.17). There are three points, namely A, B, and C, where the drift becomes zero. Point A corresponds to a stable point as any increase in n leads to a negative drift (as $(M - n)q_a < G(n)e^{-G(n)}$) that decreases state n, vice versa. On the other hand, point B is an unstable equilibrium, where any deviation will cause the system to move away from this equilibrium. The last point, C, is a stable point. However, the system suffers from a low throughput due to a large number of transmitting nodes.

It is noteworthy that the number of equilibrium points varies depending on the values of parameters, e.g. q_r and q_a. For example, if q_a is sufficiently low, there can be only one stable point.

4.4 Analysis with an Infinite Number of Users

In Section 4.3, it has been assumed that the number of users is finite. In this section, we consider the case that the number of users is infinite and assume that the number of active users follows a Poisson distribution.

4.4.1 Constant Retransmission Probability

Let X_t denote the number of active users that have new packets to transmit in time slot t in a single-channel S-ALOHA system. We assume that the probability that a user with backlogged packets retransmits a packet with probability $q = q_r$.

Provided that there are N_t backlogged packets, the number of retransmitted packets becomes

$$B_t \sim \text{Bin}(N_t, q). \tag{4.18}$$

Thus, the number of transmitted packets becomes $X_t + B_t$ in time slot t. For convenience, let $F_t \in \{0, 1, c\}$ be the feedback from the receiver, where $F_t = 0$ if there is no transmission attempt, $F_t = 1$ if there is only one transmitted packet, and $F_t = c$ if there are multiple transmitted packets. Certainly, if $F_t = 1$, the receiver is able to decode a packet. As a result, N_{t+1}, is given by

$$N_{t+1} = N_t + X_t - \mathbb{1}(F_t = 1). \tag{4.19}$$

This shows that N_t is a Markov chain with a countably infinite state space (as $N_t \in \{0, 1, \dots\}$.

Let $N_t = n$. Then, the conditional expectation of the change in the number of backlogged packets for given $N_t = n$, which is the drift of backlog, is given by

$$\mathbb{E}[N_{t+1} - N_t \mid N_t = n] = \lambda - \Pr(F_t = 1 \mid N_t = n), \tag{4.20}$$

where $\lambda = \mathbb{E}[X_t]$ is the average number of new packets. The probability that $F_t = 1$ is the sum of the following two probabilities: (i) the probability that $X_t = 0$ and $B_t = 1$ and (ii) the probability that $X_t = 1$ and $B_t = 0$. We assume that X_t follows a Poisson distribution. Then, we have

$$\Pr(X_t = 0, B_t = 1) = e^{-\lambda} \binom{n}{1} q^1 (1-q)^{n-1},$$

$$\Pr(X_t = 1, B_t = 0) = \lambda\, e^{-\lambda} \binom{n}{0} q^0 (1-q)^n. \tag{4.21}$$

Thus, the drift is given by

$$\mathbb{E}[N_{t+1} - N_t \mid N_t = n] = \lambda - e^{-\lambda} nq(1-q)^{n-1} - \lambda\, e^{-\lambda}(1-q)^n$$

$$= \lambda - e^{-\lambda}(1-q)^{n-1}(nq + \lambda(1-q)). \tag{4.22}$$

As $n \to \infty$, for any $q > 0$, we can see that $(1 - q)^{n-1} \to 0$. Thus, the drift becomes positive for a large n. In other words, the number of backlogged packets will increase once it is sufficiently large and the Markov chain, $N_t \in \{0, 1, \ldots\}$, is transient. As a result, unlike the case of a finite number of users in Section 4.3, the system becomes unstable for any $q > 0$, because all users will eventually have a growing number of backlogged packets.

Note that for multichannel ALOHA, we can see the same result as it can be seen as a collection of L independent S-ALOHA systems.

4.4.2 Adaptive Retransmission Probability

For a fixed $q > 0$, it is inevitable to avoid a large number of retransmitted packets (once N_t is sufficiently large), which results in collision and then the increase of the number of backlogged packets. To avoid this problem, the probability of retransmission, q, can be adaptively decided.

In (4.22), suppose that n is sufficiently large. Then, the second term on the RHS becomes

$$e^{-\lambda}(1 - q)^{n-1}(nq + \lambda(1 - q)) \approx e^{-\lambda} e^{-nq} nq. \tag{4.23}$$

Thus, with $q = \frac{1}{n}$, we can approximately maximize the second term (i.e. $e^{-nq} nq$) so that the drift can be minimized. The resulting drift with $q = \frac{1}{n}$ becomes

$$\mathbb{E}[N_{t+1} - N_t \mid N_t = n] \approx \lambda - e^{-\lambda} e^{-1} \left(1 + \lambda \frac{n-1}{n}\right)$$

$$\approx \lambda - (1 + \lambda)e^{-(1+\lambda)}. \tag{4.24}$$

Since the maximum of $x e^{-x}$ is e^{-1}, (4.24) implies that it is necessary to keep $\lambda < e^{-1}$ so that the drift can be negative (i.e. $\lambda < e^{-1}$ is a necessary condition for negative drift). Otherwise (i.e. if $\lambda \geq e^{-1}$), the drift becomes nonnegative and the system becomes unstable.

For $\lambda \geq 0$, it can be shown that $(1 + \lambda)e^{-(1+\lambda)}$ is a decreasing function of λ. Thus, the RHS in (4.24), i.e. $\lambda - (1 + \lambda)e^{-(1+\lambda)}$ becomes an increasing function and has one zero at $\lambda = 0.35$. In order to see this clearly, in Figure 4.6, the two functions, $y_1(\lambda) = \lambda$ and $y_2(\lambda) = (1 + \lambda)e^{-(1+\lambda)}$ are shown. For $\lambda < 0.35$, $y_1(\lambda) < y_2(\lambda)$ (and vice versa). Thus, for a negative drift, λ should be less than 0.35.

In order to see the impact of λ on the average number of backlogged packets, $\mathbb{E}[N_t]$, simulations are carried out with $q = \min\{1, \frac{1}{N_t}\}$ and $N_0 = 0$. For the mean of N_t, 100 realizations of N_t are averaged. The results are shown in Figure 4.7 when $\lambda \in \{0.34, 0.36\}$. It is shown that $\mathbb{E}[N_t]$ increases when $\lambda = 0.36$. On the other hand, when $\lambda = 0.34$, it seems that $\mathbb{E}[N_t]$ approaches a constant. This demonstrates that λ should also be controlled to avoid unstable ALOHA through access control.

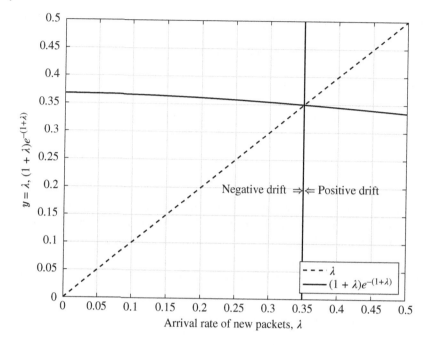

Figure 4.6 Two functions for the drift for a large n with $q = \frac{1}{n}$: $y = \lambda$ and $y = (1 + \lambda)e^{-(1+\lambda)}$.

It is noteworthy that the drift can only be negative for a large N_t if the retransmission probability, q, is determined to be inversely proportional to N_t, although $\lambda < 0.35$. Since the *total* number of the backlogged packets, N_t, is unknown to individual users, the receiver may infer and broadcast it to all the users so that they can decide q accordingly.

Alternatively, each user can attempt to estimate N_t based on the feedback. For example, if $F_t = 0$ (i.e. the channel is idle), the estimate of N_t, denoted by J_t at a user, can decrease. On the other hand, if $F_t = c$ (i.e. there are multiple packets being transmitted simultaneously), J_t can increase. As a result, the estimate can be updated as follows:

$$J_{t+1} = \max\{1, J_t + w_0 \mathbb{1}(F_t = 0) + w_1 \mathbb{1}(F_t = 1) + w_c \mathbb{1}(F_t = c)\},$$

where $w_0 < 0$ and $w_c > 0$. With the estimate of N_t, i.e. J_t, the retransmission probability becomes $q = \frac{1}{J_t}$. The coefficients, w_0, w_1, and w_c, can be determined to keep the drift negative for a large N_t. In Kelly and Yudovina (2014), the drift analysis of the Markov chain (J_t, N_t) can be found.

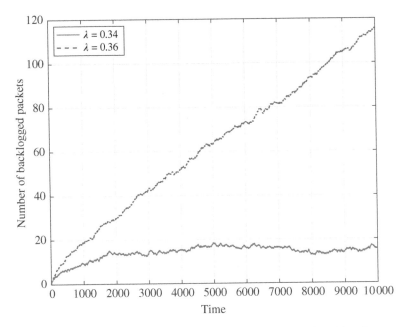

Figure 4.7 The average number of backlogged packets, $\mathbb{E}[N_t]$, by taking the average of 100 realizations $q = \min\left\{1, \frac{1}{N_t}\right\}$.

4.5 Fast Retrial

In ALOHA, as shown above, the retransmission probability plays a key role in stabilizing the system and should not be 1 to avoid consecutive collisions. There are various backoff algorithms as collision resolution mechanisms where collided users may wait random backoff times before retransmissions. However, if there are multiple channels, a user experiencing a collision may immediately attempt to send the collided packet to another channel, hoping that no other users choose the same channel. In this case, the retransmission probability can be 1.

In multichannel ALOHA, fast retrial is a simple retransmission strategy where a user with a collided packet can retransmit the collided packet immediately in the next time slot through a randomly selected channel. An example is shown in Figure 4.8 with four channels (i.e. $L = 4$) and three users. In slot t, suppose that users 1 and 3 choose channel 1, which leads to collision. In the next time slot, user 1 chooses channel 2 and user 3 chooses channel 4, while a new user, i.e. user 3, chooses channel 1. Since all users choose different channels, they can succeed to transmit their packets without collision in time slot $t + 1$. Clearly, unlike single-channel S-ALOHA, immediate retransmissions in multichannel ALOHA do not necessarily result in subsequent packet collisions.

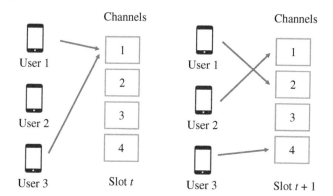

Figure 4.8 An illustration of fast retrial for multichannel ALOHA with four channels (i.e. $L = 4$) and three users.

In general, although there is no throughput gain compared to conventional multichannel ALOHA, it is known that fast retrial can lower access delay when the traffic intensity is low (Choi et al., 2006; Mutairi et al., 2013).

4.6 Multiuser Detection

In this section, we discuss the use of multiuser detection in random access when DS-CDMA is employed for multiple access, which leads to the notion of compressive random access (CRA).

4.6.1 Compressive Random Access

In this subsection, we apply DS-CDMA (direct sequence code division multiple access) to random access and discuss the notion of compressive sensing (CS) (Donoho, 2006; Candes et al., 2006) to detect the active users to exploit sparse activity.

Recall that c_k represents the spreading or signature sequence of user k in DS-CDMA (see Section 3.4), and the number of users, K, is fixed and expected to be smaller than or equal to the length of sequence, N_{pg}. In this subsection, we consider a slightly different case where a total number of users, which is now denoted by M, can be larger than N_{pg} and a fraction of users becomes active at a time. The number of *active* users is now denoted by K.

In order to differentiate the spreading sequences in conventional DS-CDMA, let g_m denote the mth signature sequence or vector (rather than c_m) for user m and let $G = [g_1 \cdots g_M]$, where M is the number of signature vectors. Let N_{pg} denote the length of signature sequences, where $N_{pg} < M$. Then, G becomes an $N_{pg} \times M$ matrix that has more columns than rows. Assume that each signature sequence is unique. In the context of CS, G is called the measurement matrix. At a receiver, the received signal is given by

$$y = \sum_{m=1}^{M} g_m s_m + n$$
$$= Gs + n, \tag{4.25}$$

where $s = [s_1 \cdots s_M]^T$ represents the signal vector and n is the background noise vector. Here, $s_m = x_m b_m$ denotes the mth element of s, where x_m and $b_m \in \{0, 1\}$ represent the data symbol of user m and the activity variable, respectively. We have $b_m = 1$ user m is active. Otherwise, it becomes zero. Due to sparse activity, s becomes a sparse vector (i.e. only a fraction of the elements are nonzero, while the others are zero).

For multiuser detection (MUD), in general, the number of rows of G should be greater than or equal to that of columns, i.e. $N_{pg} \geq M$ so that the system in (4.25) becomes an overdetermined system and the least squares solution (or any variants) can be used as an estimate of s. However, as stated earlier, M is larger than N_{pg} to support a large number of users. This means that the conventional MUD approaches in Section 3.4.2 cannot be directly applied.

It is noteworthy that in (4.25), s is sparse. Thus, if the number of active users, K, is less than or equal to N_{pg} and the index set of active users is known, the system in (4.25) can be reduced to an overdetermined system although M is larger than N_{pg}, and then any MUD approach (in Section 3.4.2; see also Verdu, 1998; Choi, 2010) can be used to find the estimates of nonzero elements of s. In other words, the detection of signals transmitted by active users can be carried out by two steps: (i) Step 1: the receiver is to detect or identify active users among M users and (ii) Step 2: MUD is carried out with active users. From this, we can see that it is important to ensure that the number of active users is sufficiently small and their indices should be known.

Finding the active users' indices is referred to as the active user detection (AUD), which is to identify or detect nonzero elements of s from y. In fact, AUD is one of sparse signal recovery problems in CS, and a number of approaches have been proposed to estimate sparse signals in CS. The reader is referred to Eldar and Kutyniok (2012) and Foucart and Rauhut (2013) for details of algorithms and recovery conditions.

4.6.2 Throughput Analysis

In (4.25), the received signal can be extended for multiple data symbols when a block or packet of data is transmitted. In this case, the tth received signal at the receiver is given by

$$\mathbf{y}_t = \mathbf{G}\mathbf{s}_t + \mathbf{n}_t, \quad t = 0, \ldots, T - 1, \tag{4.26}$$

where T represents the length of data packet, $\mathbf{s}_t = [s_{1,t} \ \cdots \ s_{M,t}]^T$ is the signal vector transmitted over time t, and $\mathbf{n}_t$ is the background noise. It is important to note that the signature matrix, $\mathbf{G}$, is invariant. Then, we can show that

$$\begin{aligned}
\mathbf{Y} &= [\mathbf{y}_0 \ \cdots \ \mathbf{y}_{T-1}] \\
&= \mathbf{G}\mathbf{S} + \mathbf{N} \in \mathbb{R}^{N_{\mathrm{pg}} \times T},
\end{aligned} \tag{4.27}$$

where $\mathbf{S} = [\mathbf{s}_0 \ \cdots \ \mathbf{s}_{T-1}] \in \mathbb{R}^{M \times T}$ and $\mathbf{N} = [\mathbf{n}_0 \ \cdots \ \mathbf{n}_{T-1}] \in \mathbb{R}^{N_{\mathrm{pg}} \times T}$. Among M users, there are only K active users transmitting signals over T-symbol duration using unique signature sequences. Thus, only K rows of $\mathbf{S}$ are nonzero. Finding the nonzero rows is a sparse multiple measurement vector (MMV) problem in the context of CS (Chen and Huo, 2006; Davies and Eldar, 2012). Under mild conditions on $\mathbf{G}$ with $T \geq N_{\mathrm{pg}}$ and a sufficiently high SNR, it is shown that up to $K = N_{\mathrm{pg}} - 1$ nonzero signals can be recovered from $\mathbf{Y}$.

To find the throughput of CRA, as mentioned earlier, we assume that each user has a unique signature sequence, $\mathbf{g}_m$. Thus, there is no packet collision. However, if the number of active users, K, is greater than $N_{\mathrm{pg}} - 1$, the receiver is unable to detect them. Thus, based on the MMV formulation, the throughput of CRA can be given by

$$\eta_{\mathrm{CRA}}(N_{\mathrm{pg}}) = \mathbb{E}[K \mathbb{1}(K \leq N_{\mathrm{pg}} - 1)], \tag{4.28}$$

where the expectation is carried out over K. Suppose each user becomes active with probability p_{a}, which is the access probability. Then, it can be shown that

$$\begin{aligned}
\eta_{\mathrm{CRA}}(N_{\mathrm{pg}}) &= \sum_{k=0}^{N_{\mathrm{pg}}-1} k \binom{M}{k} p_{\mathrm{a}}^k (1 - p_{\mathrm{a}})^{M-k} \\
&= M p_{\mathrm{a}} \sum_{k=1}^{N_{\mathrm{pg}}-1} \binom{M-1}{k-1} p_{\mathrm{a}}^{k-1} (1 - p_{\mathrm{a}})^{M-k} \\
&= M p_{\mathrm{a}} \sum_{k=0}^{N_{\mathrm{pg}}-2} \binom{M-1}{k} p_{\mathrm{a}}^k (1 - p_{\mathrm{a}})^{(M-1)-k}.
\end{aligned} \tag{4.29}$$

Let $\lambda = p_{\mathrm{a}} M$. Furthermore, consider the case that M is sufficiently large while λ is constant. Then, as $M \to \infty$, we have (Mitzenmacher and Upfal, 2005)

$$\sum_{k=0}^{N_{\mathrm{pg}}-2} \binom{M-1}{k} p_{\mathrm{a}}^k (1 - p_{\mathrm{a}})^{(M-1)-k} \to \sum_{k=0}^{N_{\mathrm{pg}}-2} \frac{\lambda^k e^{-\lambda}}{k!}. \tag{4.30}$$

Substituting (4.30) into (4.29), the throughput of CRA can be expressed as the following function of λ:

$$\eta_{\text{CRA}}(N_{\text{pg}}, \lambda) = \lambda \sum_{k=0}^{N_{\text{pg}}-2} \frac{\lambda^k e^{-\lambda}}{k!}, \tag{4.31}$$

and the maximum throughput of CRA can be obtained as follows:

$$\bar{\eta}_{\text{CRA}}(N_{\text{pg}}) = \max_{\lambda \geq 0} \lambda \sum_{k=0}^{N_{\text{pg}}-2} \frac{\lambda^k e^{-\lambda}}{k!}. \tag{4.32}$$

For comparison, we can consider multichannel ALOHA. In multichannel ALOHA, orthogonal channels are considered using an OMA scheme (e.g. FDMA or TDMA).

Figure 4.9 shows the throughput curves of CRA and multichannel ALOHA as functions of the average number of active users per slot, $G = \lambda$ when $N_{\text{pg}} = 50$. As long as λ is less than N_{pg}, we can see that CRA has a higher throughput than multichannel ALOHA. On the other hand, if λ is larger than N_{pg}, the throughput of

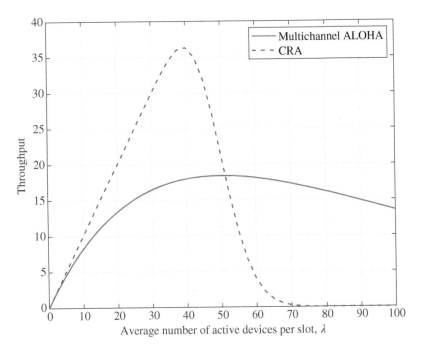

Figure 4.9 Throughput curves of CRA and multichannel ALOHA as functions of the average number of active users per slot, $G = \lambda$ when $N_{\text{pg}} = 50$.

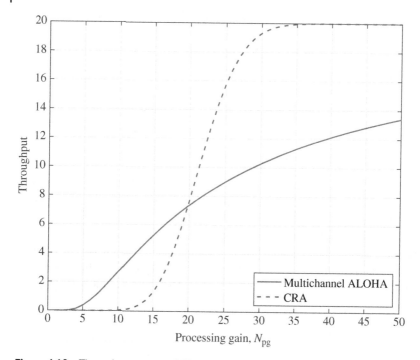

Figure 4.10 Throughput curves of CRA and multichannel ALOHA as functions of N_{pg} when the average number of active users per slot is $\lambda = 20$.

multichannel ALOHA is higher than that of CRA. This results from the assumption that CRA fails to decode any signals if the number of active users is greater than $N_{pg} - 1$ as shown in (4.28).

For a fixed λ, the impact of N_{pg} on throughput is shown in Figure 4.10 when $\lambda = 20$. As discussed above, CRA cannot perform well if $N_{pg} < \lambda$. However, CRA outperforms multichannel ALOHA as long as N_{pg} is greater than λ. In practice, for stability reasons, it is expected to run random access protocols in low-traffic environments. Thus, λ needs to be smaller than the processing gain, N_{pg}, where CRA outperforms multichannel ALOHA as shown in Figures 4.9 and 4.10.

While CRA can provide a higher throughput than multichannel ALOHA, there are a few disadvantages over multichannel ALOHA. For example, since each user must have a unique signature code that should also be known to the receiver in CRA, there is usually a limited number of users and all users must be registered before communication. In addition, since the receiver is to perform MUD as well as sparse signal recovery in CRA, it should have high enough computing power.

4.7 Further Reading

In Bertsekas and Gallager (1987), other random access protocols with their retransmission strategies can be found. The reader interested in stability analysis of ALOHA protocols is referred to Kelly and Yudovina (2014) (the drift analysis in this chapter for the case of a finite number of users is based on Bertsekas and Gallager (1987) and for the case of an infinite number of users (Kelly and Yudovina, 2014). Fast retrial was introduced in Choi et al. (2006) and further studied in Mutairi et al. (2013) and Choi (2019a).

Multipacket reception or MUD in random access was considered in Ghez et al. (1989), and then extensively studied under various names, e.g. user activity detection (Zhu and Giannakis, 2011) and CRA (Wunder et al., 2015; Choi, 2018b; see also Applebaum et al., 2012), for machine-type communication.

5

NOMA-Based Random Access

In this chapter, we discuss how non-orthogonal multiple access (NOMA) can be applied to random access to increase throughput by generating multiple channels in the power domain. In addition, an application of NOMA to Internet-of-Things (IoT) networks is presented, in which heterogeneous sensors and devices in terms of their capabilities co-exist. In particular, we demonstrate that devices with better capabilities can use NOMA in an opportunistic way to effectively utilize radio resources. Finally, NOMA is applied to a random access scheme based on code division multiple access (CDMA).

5.1 NOMA to Random Access

In order to apply NOMA to random access, it is necessary to see how NOMA is capable of generating multiple channels in the power domain. For example, consider uplink NOMA with two users as discussed in Section 3.2.2. As shown in (3.27), two users can succeed to transmit their signals as long as the transmit powers are decided to satisfy the required signal-to-interference-plus-noise ratio (SINR). Provided that their channel gains are the same (i.e. $|h_1|^2 = |h_2|^2$), we can see that $P_2 > P_1$ as $P_2 = (1 + \Gamma)P_1$. In random access, there can be more than two users, and any active user can choose one of two transmit power levels: P_1 and P_2. This is the same as the case in multichannel ALOHA with $L = 2$ channels, where any active user can choose one of two channels. From this point of view, it can be seen that NOMA is a method of generating multiple channels for random access. The following subsections detail how to apply NOMA to random access.

5.1.1 S-ALOHA with NOMA

Suppose that there are multiple users and one receiver, which is a base station (BS) for uplink transmissions, in an S-ALOHA system. In Chapter 3, we employed

Massive Connectivity: Non-Orthogonal Multiple Access to High Performance Random Access,
First Edition. Jinho Choi.

a simple model for successful transmission, in which it is assumed that a user succeeds to transmit a packet if this user is only user transmitting a packet in a given slot. We need to extend this model further to apply NOMA to S-ALOHA. Let P_k be the transmit power of user k and σ^2 the noise variance. We also assume the additive white Gaussian noise (AWGN) channel where the channel gains of users are normalized (i.e. $|h_k|^2 = 1$ for all k). Then, if user k is only a user transmitting in a slot, the signal-to-noise ratio (SNR) at the receiver is given by

$$\text{SNR} = \frac{P_k}{\sigma^2}. \tag{5.1}$$

Let $\Gamma = \frac{P^*}{\sigma^2}$ be the SNR that allows the receiver to successfully decode the transmitted packet in a given slot, where P^* represents the minimum transmit power that meets the SNR threshold, Γ. As discussed in Chapter 2, if the code rate of a given packet is less than the channel capacity, the packet can be successfully decoded (with a high probability). Thus, the code rate is to be less than $\log_2(1 + \Gamma)$ or Γ is to be greater than $2^{R_{\text{code}}} - 1$, where R_{code} represents the code rate.

In order to reproduce the collision model in S-ALOHA, let $P_k \in \{0, P\}$, where $P \geq P^*$ so that $\frac{P}{\sigma^2} \geq \Gamma$. In this setting, if $P_k = P$, then user k becomes active and transmits a packet. On the other hand, if $P_k = 0$, user k is inactive and transmits no packet. For simplicity, we assume $P = P^*$.

Suppose that there are two active users, say users 1 and 2, and the receiver is to decode the packet transmitted by user 1. Then, the resulting SNR becomes

$$\text{SNR} = \frac{P_1}{P_2 + \sigma^2} = \frac{P}{P + \sigma^2} < \Gamma, \tag{5.2}$$

which means that the receiver is unable to decode the packet transmitted by user 1. Actually, the receiver fails to decode any packet due to the interfering signal from the other user. This case is equivalent to packet collision. That is, if multiple users are active, the resulting SNR is less than the threshold SNR and the receiver fails to decode any packet. Consequently, the probability that the receiver is able to decode a packet successfully becomes (4.2) provided that the number of active users follows a Poisson distribution with mean G.

To apply the notion of power-domain NOMA to S-ALOHA, consider three different power levels as follows:

$$P_k \in \{0, P_L, P_H\}, \tag{5.3}$$

where P_H and P_L represent the high and low transmit power levels, respectively. Let $P_L = P$ and

$$P_H = \Gamma(P_L + \sigma^2).$$

Then, we can show that $\frac{P_H}{P_L + \sigma^2} = \Gamma$. Suppose that one of two nonzero power levels, i.e. P_H and P_L, can be equally likely chosen by an active user. Certainly, the receiver is able to decode a packet if there is one active user (as in S-ALOHA) regardless of the selection of transmit power level. Due to NOMA, the receiver can also decode up to two packets if two users are active simultaneously and each one chooses a different power level. To see this, suppose that the BS is to decode the signal associated with P_H, while the signal associated with P_L is regarded as the interfering signal. Since $\frac{P_H}{P_L + \sigma^2} = \Gamma$, the BS is able to decode the signal associated with P_H and remove it from the received signal using successive interference cancellation (SIC). Then, the SNR becomes $\frac{P_L}{\sigma^2} = \frac{P}{\sigma^2} = \Gamma$, which means that the BS can also decode the signal associated with P_L.

For example, consider Figure 5.1 where there are three users. Users 1 and 3 become active simultaneously and transmit their packets. In conventional ALOHA, the two packets are collided and the BS is unable to decode them. However, if the transmit power level of user 1 is P_H, while that of user 3 is P_L, then the BS can decode the packet transmitted from user 1 first and then remove it. Once the packet from user 1 is removed, the BS can decode the packet from user 3. This means that the throughput of S-ALOHA can be doubled when NOMA is applied (as will be shown below, the throughput increases by 1.6-time, not by a factor of 2).

In order to see the throughput improvement by NOMA, suppose that the number of active users follows a Poisson distribution with mean G. Then, the throughput becomes

$$\eta_{\text{NOMA}} = \Pr(\text{one active user}) + \underbrace{\Pr(\text{two active users})}_{(a)} \underbrace{\frac{1}{2}}_{} \underbrace{2}_{(b)}$$

$$= G e^{-G} + \frac{G^2}{2!} e^{-G}, \tag{5.4}$$

where (a) is the probability that one active user chooses P_H and the other active user chooses P_L and (b) is the number of successfully received packets, which is 2 as one transmits a packet with a transmit power of P_H and the other P_L. By

Figure 5.1 NOMA-ALOHA with two active users of different power levels.

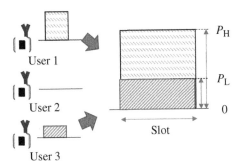

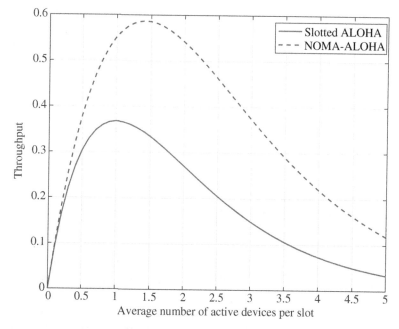

Figure 5.2 Throughput curves of S-ALOHA and NOMA-ALOHA protocols as functions of G.

comparing (4.2) and (5.4), we can see that NOMA can increase the throughput of S-ALOHA. The resulting system is referred to as NOMA-ALOHA.

Figure 5.2 shows the throughput curves of S-ALOHA and NOMA-ALOHA. Clearly, NOMA-ALOHA performs better than S-ALOHA in terms of throughput. From (5.4), it can be seen that the throughput of NOMA-ALOHA is maximized when $G = \sqrt{2}$ and the maximum throughput becomes

$$\max \eta_{\text{NOMA}} = (1 + \sqrt{2})e^{-\sqrt{2}} \approx 0.5869.$$

This shows that the maximum throughput of NOMA-ALOHA is about 1.6-time higher than that of S-ALOHA (which is $e^{-1} \approx 0.3679$).

5.1.2 More Power Levels

In Section 5.1.1, we did not consider fading channels. In wireless communications, each user has a different channel coefficient that depends on the environment including the distance between the BS and the user. In this section, with fading channels, we consider the case that there are more than two power levels.

Throughout this section, it is assumed that the channel state information (CSI) is known to users. In time division duplexing (TDD) mode, the BS can send a beacon signal at the beginning of a time slot to synchronize uplink transmissions. This beacon signal can also be used as a pilot signal to allow each user to estimate the CSI. Due to various channel impairment (e.g. fading) and the background noise, the estimation of CSI may not be perfect. However, for simplicity, we assume that the CSI estimation is perfect. Suppose that there are predetermined Q power levels that are denoted by

$$v_1 > \cdots > v_Q > 0. \tag{5.5}$$

We now assume that an active user, say user k, can randomly choose one of the power levels, say v_q, for random access. Then, the transmission power can be decided as

$$P_k = \frac{v_q}{\alpha_k}, \tag{5.6}$$

where $\alpha_k = |h_k|^2$ is the channel (power) gain from user k to the BS, so that the received signal power becomes v_q. Assuming that the variance of the background noise is normalized, i.e. $\sigma^2 = 1$, if there are no other active users, the SNR at the BS becomes v_q.

Suppose that each power level in (5.5) is decided as follows:

$$v_q = \Gamma(V_q + 1), \tag{5.7}$$

where Γ is the target SINR and $V_q = \sum_{m=q+1}^{Q} v_m$ with $V_Q = 0$. It can be shown that

$$v_q = \Gamma(\Gamma + 1)^{Q-q}. \tag{5.8}$$

These multiple power levels, $\{v_1, \ldots, v_Q\}$, can be viewed as multiple channels that can be randomly selected by active users when NOMA is applied to random access.

Example 5.1 Let $Q = 4$ and $\Gamma = 2$. From (5.8), we have

$$v_4 = \Gamma = 2,$$
$$v_3 = \Gamma(1 + \Gamma) = 6,$$
$$v_2 = \Gamma(1 + \Gamma)^2 = 18,$$
$$v_1 = \Gamma(1 + \Gamma)^3 = 54.$$

If there exists at most one active user at each power level, all the signals from active users can be decoded. To see this, suppose that there is one active user at each power level. Then, the SINR for the active user who chooses v_1 becomes $\frac{v_1}{V_1+1} = \frac{v_1}{v_2+v_3+v_4+1} = \frac{54}{27} = 2$, which is $\Gamma = 2$, and the SINRs of the other active users are also $\Gamma = 2$. For another example, suppose that there is no active user at the third power level. Then, the SINR for the active user who chooses v_1 is $\frac{v_1}{v_2+v_4+1} = \frac{54}{21}$, which is greater than Γ, and the SINRs of the other active users are greater than or equal to Γ.

With the random access scheme based on NOMA, the BS can successfully decode all signals from M active users if $M \leq Q$ and different powers are chosen. However, if there are multiple active users who choose the same power level, the signals cannot be decoded. For convenience, the event that multiple active users choose the same power level is referred to as power collision. Unlike conventional multichannel random access schemes, the power collision at each power level is not an independent event. That is, if a power collision happens at level q, the signals at levels $q + 1, \ldots, Q$ cannot be decoded, while the signals in the signals at higher power levels i.e., levels $1, \ldots, q - 1$, can be decoded if there is no power collision.

Example 5.2 Consider the case in Example 5.1, where $Q = 4$ and $\Gamma = 2$. Let $M = 3$ and assume that one user chooses v_1 and the other two users choose v_4. Then, the signal from the user choosing v_1 can be decoded as the SINR becomes $\frac{v_1}{2v_4+1} = \frac{54}{5}$, which is greater than $\Gamma = 2$, although the signals from the other two users cannot be decoded (as their SINRs are $\frac{v_4}{v_4+1} = \frac{2}{3}$) due to the power collision.

For the performance of random access based on NOMA, we consider the conditional throughput that is the average number of signals that are successfully decoded for given M. A bound on the (conditional) throughput can be found as follows.

The conditional throughput for given M ($M \leq Q$) active users, denoted by $\eta(M; Q)$, is bounded as

$$\eta(M; Q) \geq \underline{\eta}(M; Q)$$
$$= M \prod_{m=1}^{M-1} \left(1 - \frac{m}{Q}\right). \tag{5.9}$$

To obtain (5.9), consider the throughput that corresponds to the case that all M signals can be decoded. Since the probability that all M signals have different power levels is $\prod_{m=1}^{M-1}\left(1 - \frac{m}{Q}\right)$ (Mitzenmacher and Upfal, 2005), we can have (5.9). As illustrated in Example 5.2, since it is also possible to decode some signals in the presence of power collisions, (5.9) becomes a lower-bound.

In the case of $M = 2$, the BS can decode two signals if two active users choose different power levels. If they choose the same power level, no signal can be decoded. Thus, the lower-bound in (5.9) becomes exact. Note that $\eta(M; Q) \geq 0$ for any value of M. Thus, the lower-bound in (5.9) is valid for any value of M as $\underline{\eta}(M; Q) = 0$ for $M > Q$.

From (5.9), we can show that the resulting random access scheme based on NOMA can have a higher throughput as Q increases. However, from (5.8), since the highest power level, $v_1 = \Gamma(\Gamma + 1)^{Q-1}$, grows exponentially with Q, a large Q

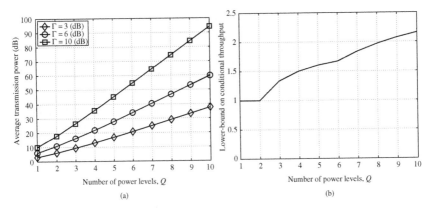

Figure 5.3 Performance of NOMA for different numbers of power levels, Q: (a) the average transmission power; (b) the lower-bound on conditional throughput (the optimal value of M is chosen with the lower-bound in (5.9)).

becomes impractical. In Figure 5.3, we illustrate the average transmission power of NOMA for different values of Q and target SINR, Γ. Figure 5.3a shows the average transmission power of NOMA for different numbers of power levels, Q. We can see that the increase of the average transmission power is significantly higher as Q increases. On the other hand, the increase of the throughput with Q is not significant as shown in Figure 5.3b. Thus, it may not be desirable to have a large Q. Note that the conditional throughput in Figure 5.3b is the lower-bound in (5.9) where the optimal value of M is chosen to maximize the bound, i.e.

$$\max_{1 \le M \le Q} \eta(M; Q).$$

In the following example, we discuss another approach that can be used to find a lower-bound on the throughput of NOMA-ALOHA.

Example 5.3 Suppose that $Q = 3$ and assume that the receiver fails to decode any signal due to power collisions. To find the throughput, we can formulate a balls-into-bins problem. If there are two active users, we have two balls and $Q = 3$ bins. In addition, let X_q be the number of users choosing bin q. Then, the event that two signals can be decoded is $(X_1, X_2, X_3) = (1, 1, 0), (0, 1, 1)$, or $(1, 0, 1)$. Using the multinomial distribution, the resulting probability becomes

$$\text{Pr (decoding} \mid M = 2) = \underbrace{3}_{\text{number of cases}} \times \underbrace{\frac{2!}{1!1!0!} \frac{1}{3^2}}_{\text{probability}} = \frac{2}{3}.$$

If there are $M = 3$ active users, the event that all the signals can be decoded is $(X_1, X_2, X_3) = (1, 1, 1)$. All the other cases have power collisions (i.e. there exist events that $X_q > 1$). The probability of decoding becomes

$$\Pr(\text{decoding} \mid M = 3) = \underbrace{1}_{\text{number of cases}} \times \underbrace{\frac{3!}{1!1!1!} \frac{1}{3^3}}_{\text{probability}} = \frac{2}{9}.$$

If $M \geq 4$, there exists at least one power collision and the receiver is unable to decode signals. As a result, when M follows a Poisson distribution with mean G, the throughput becomes

$$\eta = \underbrace{G\,e^{-G}}_{=\Pr(M=1)} + 2 \times \frac{2}{3} \times \underbrace{\frac{G^2}{2!}e^{-G}}_{=\Pr(M=2)} + 3 \times \frac{2}{9} \times \underbrace{\frac{G^3}{3!}e^{-G}}_{=\Pr(M=3)}$$

$$= e^{-G}\left(G + \frac{2}{3}G^2 + \frac{1}{9}G^3\right)$$

$$= G\,e^{-G}\left(1 + \frac{G}{3}\right)^2. \tag{5.10}$$

In Figure 5.4, the throughput of NOMA-ALOHA when $Q = 3$ is shown with the lower-bound in (5.10). For simulation results, the target SINR is set to $\Gamma = 10\,\text{dB}$.

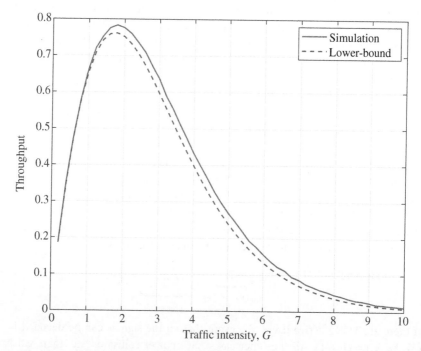

Figure 5.4 Throughput of NOMA-ALOHA when $Q = 3$.

5.2 Multichannel ALOHA with NOMA

It is straightforward to apply NOMA to multichannel ALOHA with L orthogonal subchannels, which will be studied in this section. Furthermore, we study channel-dependent selection for subchannel as well as power level to reduce the transmission power or improve the energy efficiency.

5.2.1 Multichannel ALOHA with NOMA and Throughput Analysis

Suppose that each (orthogonal) subchannel in multichannel ALOHA employs NOMA. Then, we can have a total of QL subchannels, where each subchannel has Q power levels for NOMA. To see the throughput, suppose that there are M_l active users that choose the lth subchannel. In addition, let $\mathbf{m} = [M_1 \ \cdots \ M_L]^T$ and $M = \sum_{l=1}^{L} M_l$, which is the total number of active users. Denote by $\eta_{\text{NMA}}(\mathbf{m}; Q, L)$ the conditional throughput of NOMA-ALOHA for given $\mathbf{m}$. Then, we have

$$\eta_{\text{NMA}}(\mathbf{m}; Q, L) = \sum_{l=1}^{L} \eta(M_l; Q). \tag{5.11}$$

Let

$$T_{\text{NMA}}(M; Q, L) = \mathbb{E}[\eta_{\text{NMA}}(\mathbf{m}; Q, L) \mid M], \tag{5.12}$$

which is the conditional throughput for given M, where the expectation is carried out over $\mathbf{m}$ for given M. Suppose that each active user can choose a subchannel uniformly at random. Then, it can be shown that

$$T_{\text{NMA}}(M; Q, L) = \sum_{\mathbf{m}} \eta_{\text{NMA}}(\mathbf{m}; Q, L) p(\mathbf{m} \mid M)$$

$$= \sum_{\mathbf{m}} \sum_{l=1}^{L} \eta(M_l; Q) p(\mathbf{m} \mid M), \tag{5.13}$$

where $p(\mathbf{m} \mid M)$ is the pmf of multinomial random variables that is given by $p(\mathbf{m} \mid M) = \frac{M}{M_1! \cdots M_L!} \left(\frac{1}{L} \right)^M$. By marginalization, we can show that

$$\sum_{\mathbf{m}} \sum_{l=1}^{L} \eta(M_l; Q) p(\mathbf{m}) = L \sum_{n=1}^{M} \eta(n; Q) p(n; M),$$

where

$$p(n; M) = \binom{M}{n} \left(\frac{1}{L} \right)^n \left(1 - \frac{1}{L} \right)^{M-n},$$

which is the binomial distribution with parameters M and $p = \frac{1}{L}$. Finally, the conditional throughput for given M can be found as

$$T_{\text{NMA}}(M; Q, L) = \mathbb{E}[\eta_{\text{NMA}}(\mathbf{m}; Q, L) \mid M]$$

$$= L\mathbb{E}[\eta(N, Q)], \tag{5.14}$$

where N becomes the binomial random variable with parameters M and $p = \frac{1}{L}$. This shows that, as with multichannel ALOHA, throughput increases linearly with the number of subchannels, L.

Suppose that there are a large number of users, while a fraction of them are active at a time. Then, the number of active users follows a Poisson distribution. Using this Poisson approximation, from (5.9) and (5.14), a lower-bound on the average throughput can be found as

$$
\begin{aligned}
T_{\text{NMA}}(Q, L) &= \mathbb{E}[T_{\text{NMA}}(M; Q, L)] \\
&\geq L \sum_{q=1}^{Q} \underline{\eta}(q; Q) p_\lambda(q) \\
&= L \sum_{q=1}^{Q} q \left(\prod_{m=1}^{q-1} \left(1 - \frac{m}{Q}\right) \right) \frac{e^{-\lambda} \lambda^q}{q!},
\end{aligned}
\tag{5.15}
$$

where $\lambda = p_a N$. Here, N is the total number of users and p_a is the access probability, i.e. the probability that a user becomes active.

If $Q = 2$, as mentioned earlier, the lower-bound is exact (because $M \leq Q$). Thus, we can show that

$$
T_{\text{NMA}}(2, L) = L \left(e^{-\lambda} \lambda + \frac{e^{-\lambda} \lambda^2}{2!} \right),
$$

which is identical to (5.4) with $G = \lambda$ and $L = 1$.

In addition, if $Q = 1$, NOMA-ALOHA is reduced to standard multichannel ALOHA that has the following throughput:

$$
T_{\text{NMA}}(1, L) = T_{\text{MA}}(L) = L \lambda \, e^{-\lambda}.
$$

From this, we can see that the maximum throughput of NOMA-ALOHA with $Q = 2$ is 1.6 times higher than that of standard multichannel ALOHA. Furthermore, as $\eta(n; Q)$ increases with Q, the lower-bound on the average throughput increases with Q. Consequently, we can see that NOMA-ALOHA can improve the throughput of multichannel ALOHA without any bandwidth expansion based on the notion of NOMA.

Figure 5.5 shows the throughput of NOMA-ALOHA for different numbers of subchannels, L, when $N = 200$, $p_a = 0.05$, and $Q \in \{1, 4\}$. The lower-bound is obtained from (5.15). As expected, we can observe that the throughput increases with the number of subchannels, L. More importantly, we can see that the throughput of NOMA-ALOHA ($Q = 4$) is higher than that of (conventional) multichannel ALOHA ($Q = 1$). In particular, when $L = 4$, the throughput of NOMA-ALOHA becomes about four times higher than that of multichannel ALOHA, while the throughput gap decreases with L. This demonstrates that when the number of subchannels is limited in multichannel ALOHA, NOMA can help

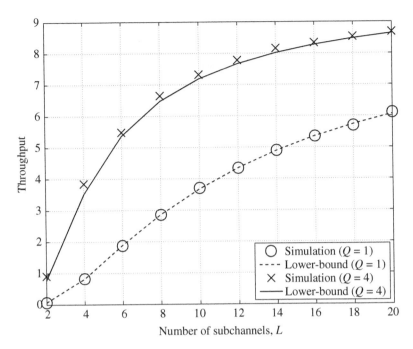

Figure 5.5 Throughput of NOMA-ALOHA for different number of subchannels, L, when $N = 200$, $p_a = 0.05$, and $Q \in \{1, 4\}$.

improve the throughput of random access. For example, NOMA-ALOHA ($Q = 4$) can achieve a throughput of 3.5 with $L = 4$, while the same throughput can be obtained by multichannel ALOHA with $L = 10$ (in this case, NOMA-ALOHA has 2.5 times higher spectral efficiency than multichannel ALOHA).

In order to see the impact of the number of power levels, Q, on the throughput of NOMA-ALOHA, we show the throughput for different values of Q in Figure 5.6 when $N = 200$, $p_a = 0.05$, and $L = 6$. As expected, the throughput increases with Q without any bandwidth expansion (i.e. with a fixed L). For example, with $Q = 4$, the throughput can be three times higher than that of (conventional) multichannel ALOHA (i.e. NOMA-ALOHA with $Q = 1$). However, the improvement of throughput becomes limited when L is sufficiently large.

However, the increase of Q results in the increase of transmission power. To mitigate the increase of transmission power, we can consider a channel-dependent subchannel/power-level selection scheme in the following subsection.

5.2.2 Channel-Dependent Selection

In above, we assume that each active user chooses a subchannel and a power level independently and uniformly at random. The selection of subchannel and power

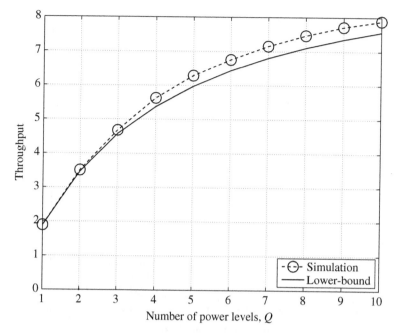

Figure 5.6 Throughput for different values of Q when $N = 200$, $p_a = 0.05$, and $L = 6$.

level can depend on the channel gain and it may result in the improvement in terms of energy efficiency (or the decrease of transmission power).

Suppose that users are uniformly distributed within a cell of radius D. We assume that the large-scale fading coefficient of user k is given by (Tse and Viswanath, 2005)

$$\mathbb{E}[\alpha_{i,k}] = \overline{\alpha}_k = A_0 d_k^{-\kappa}, \ 0 < d_k \le D, \tag{5.16}$$

where $\alpha_{i,k} = |h_{i,k}|^2$, κ is the path loss exponent, A_0 is constant, and d_k is the distance between the BS and user k. Thus, the large-scale fading coefficient depends on the distance.

For illustration purposes, suppose that $Q = 2$. According to the large-scale fading coefficients or distances, we can divide users into two groups as follows:

$$\mathcal{N}_1 = \{k \mid d_k \le \tau\},$$
$$\mathcal{N}_2 = \{k \mid d_k > \tau\}.$$

If an active user belongs to $\mathcal{N}_1$, this user selects v_1. Otherwise, the user selects v_2. That is, a user located far away from the BS tends to choose a smaller v_q to reduce the overall transmission power. We may decide the threshold value τ to satisfy the following condition:

$$\mathbb{E}[|\mathcal{N}_1|] = \mathbb{E}[|\mathcal{N}_2|] = \frac{N}{2},$$

so that each group has the same number of users on average. In this case, we have $\tau = \frac{D}{\sqrt{2}}$. Consequently, the large-scale fading coefficient is used as a random number for the power level selection and the value of τ is decided to make sure that $\Pr(k \in \mathcal{N}_q) = \frac{1}{2}$, (i.e. for a uniform power level selection at random).

The above approach can be generalized for $Q \geq 2$. To this end, let

$$\mathcal{N}_q = \{k \mid \tau_{q-1} < d_k \leq \tau_q\}. \tag{5.17}$$

Under the assumption that users are uniformly distributed in a cell of radius D, we have $\tau_0 = 0$ and $\tau_q = D\sqrt{\frac{q}{Q}}, q = 1, \ldots, Q$, to satisfy

$$\Pr(k \in \mathcal{N}_q) = \frac{1}{Q}, \quad q = 1, \ldots, Q,$$

which also results in $\mathbb{E}[|\mathcal{N}_q|] = \frac{N}{Q}$. To minimize the transmission power, an active user belongs to $\mathcal{N}_q$ chooses v_q.

Furthermore, when an active user in $\mathcal{N}_q$ chooses one of L subchannels in NOMA-ALOHA, the user may choose the subchannel of the maximum channel gain to further minimize the transmission power. As a result, the transmission power of user k can be decided as

$$P_k = \frac{v_q}{\max_i \alpha_{i,k}}, \quad k \in \mathcal{N}_q. \tag{5.18}$$

Note that in this case, if $\alpha_{1,k}, \ldots, \alpha_{L,k}$ are independent and identically distributed (iid), the selection of subchannel is carried out independently and uniformly at random. The selection scheme resulting in (5.18) is referred to as the channel-dependent subchannel/power-level selection scheme.

To see the performance of NOMA-ALOHA, we consider the channel gains in (5.16) and the following:

$$\alpha_{i,k} = \bar{\alpha}_k u_{i,k}^2, \tag{5.19}$$

where $u_{i,k}$ is an independent Rayleigh random variable with $\mathbb{E}[u_{i,k}^2] = 1$ (i.e. small-scale fading is assumed to be Rayleigh distributed). For the path loss exponent, κ, in (5.16), we assume that $\kappa = 3.5$. In addition, we assume that $D = 1$ and $A_0 = 1$ in (5.16) for normalization purposes.

To see the impact of this selection scheme on the average transmission power, we present simulation results in Figure 5.7 where the average transmission power is shown for different values of Q when $N = 200$, $p_a = 0.05$, $L = 6$, and $\Gamma = 6\,\mathrm{dB}$. Furthermore, for performance comparisons, we consider the random selection for subchannel and power level regardless of the channel conditions. Since the

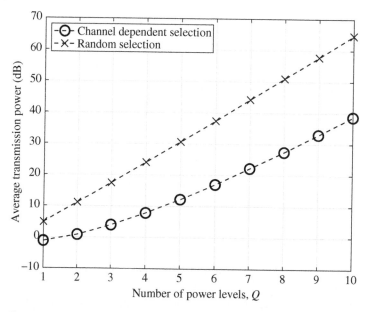

Figure 5.7 Average transmission power for different values of Q when $N = 200$, $p_a = 0.05$, $L = 6$, and $\Gamma = 6$ dB.

transmission power can be arbitrarily high due to the channel inversion power control in (5.6), we assume that the transmission power is limited to be less than or equal to $10Q$ dB (i.e. truncated power control is assumed) in simulations hereafter. The corresponding results are shown with the legend "Sim (Random)" in Figure 5.7. We can observe that the average transmission power increases with Q, while the channel-dependent selection scheme provides a much lower average transmission power than the (channel-independent) random selection scheme.

Figure 5.8 shows the average transmission power for different numbers of subchannels, L, when $N = 200$, $p_a = 0.05$, $Q = 4$, and $\Gamma = 6$ dB. As expected, the average transmission power decreases with L. On the other hand, the average transmission power with the random selection does not depend on L. Consequently, we can see that although a large L does not help improve the throughput significantly (with a fixed p_a) in NOMA-ALOHA as shown in Figure 5.5, it can be effective for improving energy efficiency with channel-dependent selection.

5.3 Opportunistic NOMA

In this section, we consider an opportunistic NOMA approach that can improve the throughput of an IoT network where multiple orthogonal resource blocks

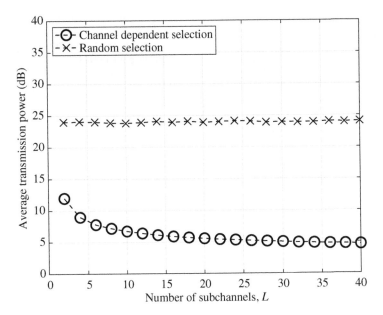

Figure 5.8 Average transmission power for different numbers L of subchannels when $N = 200$, $p_a = 0.05$, $Q = 4$, and $\Gamma = 6$ dB.

(RBs) or channels (i.e. throughout this section, the terms RBs and channels are used interchangeably) are available so that a device or user (the terms device and user are used interchangeably) can choose one of them when accessing the network. For convenience, denote by L the number of channels in the IoT network. The model in Bonnefoi et al. (2018) is considered for the IoT network and no fading is assumed in this section.

It is assumed that there are two groups of devices. One group consists of low-cost static devices (SDs) of a low duty cycle or access probability that have limited RF capabilities. As a result, we assume that each SD can only use one fixed channel that is predetermined. The other group consists of dynamic devices (DDs) that are more capable than SDs in the following ways:

1. each DD can choose a channel dynamically;
2. the transmit power of DDs is higher than that of SDs so that power-domain NOMA can be employed.

5.3.1 System Model

As mentioned earlier, each SD has a fixed channel (among L channels) to communicate with a gateway or BS. Thus, the number of SDs on channel l, denoted

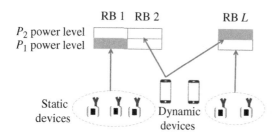

Figure 5.9 An illustration of the system model with L RBs to support both SDs and DDs using power-domain NOMA.

by S_l, is assumed to be a constant. On the other hand, each DD is more flexible and able to dynamically choose a channel. In Figure 5.9, we illustrate the system model with L RBs to support both SDs and DDs using power-domain NOMA. In addition, as mentioned above, each DD is to transmit a higher power than SDs to avoid collision with SDs by exploiting power-domain NOMA (Choi, 2017b).

Let $\mathcal{K}_l$ and $\overline{\mathcal{K}}_l$ denote the index sets of active SDs and DDs that transmit their signals to channel l, respectively. Let $K_l = |\mathcal{K}_l|$ and $\overline{K}_l = |\overline{\mathcal{K}}_l|$. The signals from the kth active SD and DDs are denoted by s_k and $\bar{s}_k$, respectively. Then, the received signal through channel l is given by

$$y_l = \sqrt{P_1} \sum_{k \in \mathcal{K}_l} s_k + \sqrt{P_2} \sum_{k \in \overline{\mathcal{K}}_l} \bar{s}_l + n_l, \tag{5.20}$$

where $n_l \sim \mathcal{N}(0, \sigma^2)$ denotes the background noise, and P_1 and P_2 represent the power levels of SDs and DDs, respectively. We assume that $\mathrm{Var}(s_k) = \mathrm{Var}(\bar{s}_k) = 1$ for normalization with $\mathbb{E}[s_k] = \mathbb{E}[\bar{s}_k] = 0$ for all k. Thus, if $|\overline{\mathcal{K}}_l| = \overline{K}_l = 1$, the SINR of DDs is given by

$$\mathrm{SINR}_2 = \frac{P_2}{K_l P_1 + \sigma^2}. \tag{5.21}$$

It is assumed that the signal from one DD through channel l is decodable if

$$\mathrm{SINR}_2 \geq \Gamma, \tag{5.22}$$

where $\Gamma > 0$ denotes the SINR threshold for successful decoding. Let

$$\Gamma_m = \frac{P_2}{m P_1 + \sigma^2}, \quad m = 1, 2, \dots, \tag{5.23}$$

which is the SINR of DD in the presence of m SDs in the same channel. Here, m can be seen as a design parameter. Suppose that there is only one DD in channel l and $\Gamma = \Gamma_m$. In this case, if there is at most m active SD in the same channel, the signal from the DD is decodable as $\mathrm{SINR}_2 \geq \Gamma$. Of course, the signal of the active SD can also be decodable if there is no SD collision after SIC.

It is noteworthy that the packet collision with multiple DDs in a channel results in decoding failure of the SD in the same channel due to error propagation. Thus, a sufficiently low probability of packet collision with DDs is desirable. In other words, the average number of active DDs, denoted by λ, has to be lower than the number of channels, L. Throughout this section, therefore, we assume that $\lambda \leq L$.

Note that the total number of SDs is $M_1 = \sum_{l=1}^{L} S_l$. Let p_1 denote the access probability of SDs that are active independently. In general, the total number of DDs, denoted by M_2, is also finite. Denote by p_2 the access probability of DDs becoming active independently. Thus, we have

$$\lambda = \mathbb{E}[N_2] = M_2 p_2, \tag{5.24}$$

where $N_2 = \sum_{l=1}^{L} \overline{K}_l$ is the number of active DDs. For convenience, with a sufficiently large M_2 and a low p_2, we assume that N_2 is a Poisson random variable.

5.3.2 Throughput Analysis

In this section, we focus on the throughput analysis in order to demonstrate the performance improvement by exploiting the notion of power-domain NOMA to support SDs and DDs in different ways.

For comparisons, we discuss the throughput of the conventional random access approach in Bonnefoi et al. (2018), where no power-domain NOMA is considered. In each channel, both SDs and DDs transmit signals with power $P = P_1 = P_2$ as NOMA is not used. Thus, provided that there are N_2 active DDs, the conditional probability that one DD can successfully transmit its packet through channel l is given by

$$\mathbb{P}_{2,l}(N_2) = \left(1 - p_1\right)^{S_l} \binom{N_2}{1} q_l \left(1 - q_l\right)^{N_2 - 1}, \tag{5.25}$$

where q_l is the probability that an active DD chooses channel l or DD's selection probability of channel l. Thus, the throughput of DDs, which is the average number of successfully transmitted packets by DDs, is given by

$$\eta_{\text{conv},2}(\mathbf{q}) = \mathbb{E}\left[\sum_{l=1}^{L} \mathbb{P}_{2,l}(N_2)\right]$$

$$= \sum_{l=1}^{L} \left(1 - p_1\right)^{S_l} \sum_{k=0}^{\infty} k q_l \left(1 - q_l\right)^{k-1} \frac{e^{-\lambda} \lambda^k}{k!}$$

$$= \sum_{l=1}^{L} \left(1 - p_1\right)^{S_l} \lambda q_l \, e^{-\lambda q_l}. \tag{5.26}$$

Let $\mathbb{P}_{1,l}(N_2)$ denote the conditional probability that an active SD in channel l can successfully transmit its packet provided that there are N_2 active DDs, which is given by

$$\mathbb{P}_{1,l}(N_2) = S_l p_1 \left(1 - p_1\right)^{S_l - 1} \left(1 - q_l\right)^{N_2}. \tag{5.27}$$

Then, the throughput of SDs can also be found as

$$
\eta_{\text{conv},1}(\mathbf{q}) = \mathbb{E}\left[\sum_{l=1}^{L} \mathbb{P}_{1,l}(N_2)\right]
$$

$$
= \sum_{l=1}^{L} S_l p_1 \left(1 - p_1\right)^{S_l - 1} e^{-\lambda q_l}. \tag{5.28}
$$

For a given channel l, the throughput of SDs per channel is a decreasing function of q_l as shown in (5.28). On the other hand, as shown in (5.26), the throughput of DDs per channel is an increasing function of q_l when $q_l \le \frac{1}{\lambda}$. The relationship between the throughput (per channel) and DD's access probability q_l is illustrated in Figure 5.10. As a result, the increase of throughput of DDs leads to the decrease of throughput of SDs.

We now consider the case that NOMA is applied. Let ω_l denote the probability that an active DD can successfully transmit its packet when it is only one active DD. With $\Gamma = \Gamma_1$, from (5.21) and (5.22), it can be given by

$$
\omega_l = \Pr\left(\text{SINR}_2 \ge \Gamma \mid \overline{K}_l = 1\right)
$$

$$
= (1 - p_1)^{S_l} + \binom{S_l}{1} p_1 (1 - p_1)^{S_l} = (1 - p_1)^{S_l} + S_l p_1 (1 - p_1)^{S_l}. \tag{5.29}
$$

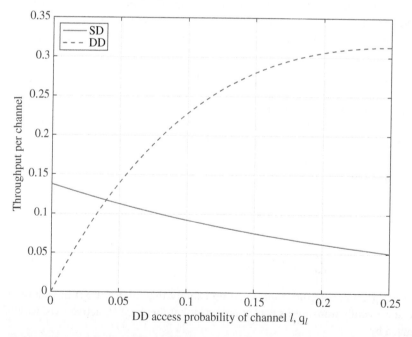

Figure 5.10 Throughput of the conventional approach per channel as a function of DD access probability, q_l, for SDs and DDs when $p_1 = 0.01$, $S_l = 16$, and $\lambda = 4$.

As shown in (5.29), ω_l only depends on the number of active SDs in channel l. Provided that there are N_2 active DDs, the conditional probability that an active DD choosing channel l can successfully transmit its packet is given by

$$\mathbb{P}_{2,l}(N_2) = \omega_l \binom{N_2}{1} q_l(1 - q_l)^{N_2-1}. \tag{5.30}$$

Thus, the throughput of DDs becomes

$$\eta_2(\mathbf{q}) = \mathbb{E}\left[\sum_{l=1}^{L} \omega_l N_2 q_l(1 - q_l)^{N_2-1}\right] = \lambda \sum_{l=1}^{L} \omega_l q_l \, e^{-q_l \lambda}. \tag{5.31}$$

To decode the signal from an active SD in a channel, it is necessary to decode any active DD and perform SIC. Thus, the throughput of SDs can be given by

$$\eta_1(\mathbf{q}) = \mathbb{E}\left[\sum_{l=1}^{L} S_l p_1(1 - p_1)^{S_l-1} \left(N_2(1 - q_l)^{N_2-1} + (1 - q_l)^{N_2}\right)\right]$$

$$= \sum_{l=1}^{L} S_l p_1(1 - p_1)^{S_l-1} e^{-q_l \lambda}(1 + \lambda q_l), \tag{5.32}$$

where the expectation is carried out over N_2.

From (5.32) and (5.28), we can see that NOMA can help improve the throughput of SDs compared with its counterpart without NOMA in Bonnefoi et al. (2018) at the cost of high transmit power of DDs. In other words, the presence of DDs has less impact on the performance of SDs when power domain NOMA is used. Furthermore, if $q_l = \frac{1}{L}$, we have

$$e^{-q_l \lambda}(1 + \lambda q_l) = e^{-\frac{\lambda}{L}}\left(1 + \frac{\lambda}{L}\right) \leq 0.7358,$$

as $\lambda \leq L$. This demonstrates that the throughput of SDs can be degraded by a factor of up to 0.7358 due to the presence of DDs. On the other hand, as shown in (5.28), without NOMA, the throughput of SDs can be degraded by a factor of up to $e^{-1} = 0.3679$ due to the presence of DDs.

The following result shows that the throughput of the NOMA-based approach is higher than that of the counterpart without NOMA:

$$\max_{\mathbf{q}} \eta_1(\mathbf{q}) \geq \max_{\mathbf{q}} \eta_{\text{conv},1}(\mathbf{q}),$$

$$\max_{\mathbf{q}} \eta_2(\mathbf{q}) \geq \max_{\mathbf{q}} \eta_{\text{conv},2}(\mathbf{q}),$$

$$\max_{\mathbf{q}} \eta_1(\mathbf{q}) + \eta_2(\mathbf{q}) \geq \max_{\mathbf{q}} \eta_{\text{conv},1}(\mathbf{q}) + \eta_{\text{conv},2}(\mathbf{q}). \tag{5.33}$$

In Figure 5.11, the relation between the throughput (per channel) and DD's selection probability q_l is illustrated with the same values of parameters as those in Figure 5.10. Comparing Figures 5.10 and 5.11, it is clear that the NOMA-based

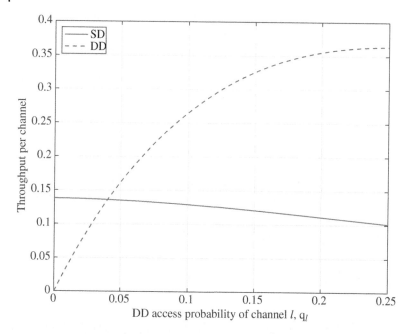

Figure 5.11 Throughput of the NOMA-based approach per channel as a function of DD selection probability, q_l, for SDs and DDs when $p_1 = 0.01$, $S_l = 16$, and $\lambda = 4$.

approach can provide a higher throughput than the conventional one (i.e. counterpart without NOMA) for both SDs and DDs.

In Figure 5.12, we also compare the conventional approach and NOMA-based approach in terms of the total throughput when λ varies from 0 to L with $L = 10$, $S_l = S = 10$, $p_1 = 0.1$, and $q_l = \frac{1}{L}$ for all l. It can be seen that the NOMA-based approach has a higher throughput than the conventional one thanks to power-domain NOMA.

Recall that Γ_m is the SINR of DD when there are m SDs in the same channel. Letting $\Gamma = \Gamma_m$, it is clear that ω_l increases with m. This results in the increase of the throughput of DDs at the cost of increasing transmit power of DDs. In Figure 5.13, we show the performance of the conventional random access approach and the NOMA-based approach in terms of throughput for various values of p_1 with SINR thresholds Γ_1 and Γ_2 when $L = 10$, $S_l = S = 10$, $\lambda = 7$ and $q_l = \frac{1}{L}$ for all l. It is demonstrated that as the SINR threshold for DDs increases, the sum throughput increases. Thus, more capable DDs can help improve the sum throughput.

As shown in (5.32), a salient feature of the NOMA-based approach is that the throughput of SDs is less dependent on q_l as long as λq_l is sufficiently low, which might be the case that $\lambda < L$. This is the expected result as DDs are opportunistic

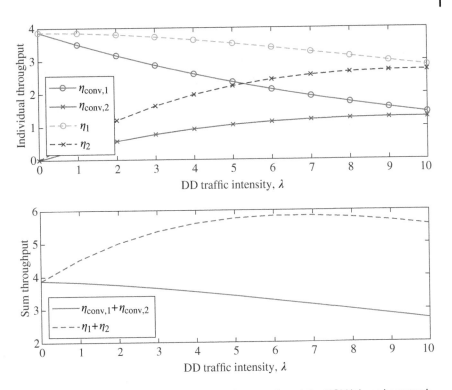

Figure 5.12 Performance of the conventional approach and the NOMA-based approach in terms of throughput for various values of λ when $L = 10$, $S_l = S = 10$, $p_1 = 0.1$ and $q_l = \frac{1}{L}$ for all l.

in accessing channels. That is, the presence of smart DDs should not have a serious impact on poorly capable SDs. Based on this, we can consider the following optimization problem:

$$\mathbf{q}^* = \arg\max_{\mathbf{q}} \eta_2(\mathbf{q})$$
$$\text{subject to } \sum_{l=1}^{L} q_l = 1, \ q_l \geq 0. \tag{5.34}$$

The solution of the problem in (5.34) is given by

$$q_l^* = \begin{cases} \dfrac{1 - \mathbb{W}\left(\dfrac{v e}{\omega_l}\right)}{\lambda}, & \text{if } v \leq \omega_l, \\ 0, & \text{otherwise,} \end{cases} \tag{5.35}$$

where $\mathbb{W}(\cdot)$ is the Lambert W function and v is the Lagrange multiplier. The value of the Lagrange multiplier has to be decided to satisfy $\sum_{l=1}^{L} q_l^* = 1$.

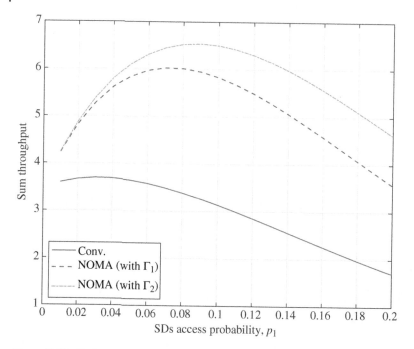

Figure 5.13 Performance of the conventional random access approach and the NOMA-based approach in terms of throughput for various values of p_1 with SINR thresholds Γ_1 and Γ_2 when $L = 10$, $S_l = S = 10$, $\lambda = 7$ and $q_l = \frac{1}{L}$ for all l.

To show (5.35), from (5.34), we can consider the following unconstrained optimization problem:

$$\min_{\mathbf{q}} \sum_{l=1}^{L} \omega_l q_l \, e^{-q_l \lambda} - v \sum_{l=1}^{L} q_l. \tag{5.36}$$

By taking the derivative with respect to q_l and setting to 0, we have

$$e^{-q_l \lambda}(1 - q_l \lambda) = \frac{v}{\omega_l}, \quad \text{for all } l. \tag{5.37}$$

Then, (5.37) is rewritten as

$$u_l \, e^{u_l} = z_l, \tag{5.38}$$

where $u_l = 1 - q_l \lambda$ and $z_l = \frac{v e}{\omega_l}$. Since $u = \mathbb{W}(z)$ when $u \, e^u = z$, we have

$$u_l = \mathbb{W}(z_l) \quad \text{or} \quad q_l = \frac{1 - \mathbb{W}(z_l)}{\lambda}. \tag{5.39}$$

Since $\mathbb{W}(z)$ is an increasing function of $z \geq 0$ and becomes 1 when $z = e$, for $z_l > e$ or $v > \omega_l$, q_l has to be 0. This results in (5.35), which completes the proof.

Note that finding v that satisfies $\sum_l q_l^* = 1$ is straightforward as $1 - \mathbb{W}(z_l)$ is a nonincreasing function of v. It can be found by any simple numerical technique, e.g. the bi-section method (Boyd and Vandenberghe, 2009).

According to (5.35), it can be seen that q_l^* increases with ω_l. This implies that DDs need to be opportunistic in choosing channels. In particular, if ω_l is lower (i.e. S_l is smaller), DDs need to transmit more packets in channel l to increase the throughput.

5.3.3 Opportunistic NOMA for Channel Selection

As discussed above, DDs can choose channels dynamically to increase the probability to successfully transmit their packets in the presence of SDs. Since the transmit power of DD is high due to NOMA, each DD can attempt channel selection to minimize the transmit power as well as increase the probability of success. In this section, we consider an approach that takes into account channel gain.

Let $G_{l;k}$ be the channel power gain from the kth active DD to the BS. In TDD mode, due to the channel reciprocity, the channel gains for uplink and downlink channels are the same. Thus, the kth active DD can estimate its channel gain from a beacon signal transmitted by the BS and find $G_{l;k}$ of channel l. Recall that P_2 is the power level for DDs. Then, the transmit power, denoted by $P_{\text{tx},k}$, can be decided to satisfy $P_2 = \frac{P_{\text{tx},k} G_{l;k}}{\sigma^2}$ as follows:

$$P_{\text{tx},k} = \frac{P_2 \sigma^2}{G_{l;k}}. \tag{5.40}$$

Thus, in order to minimize the transmit power, the kth active DD can choose the channel to transmit as follows:

$$l^* = \arg\min_l \frac{P_2 \sigma^2}{G_{l;k}} = \arg\max_l G_{l;k}. \tag{5.41}$$

In addition, if $G_{l;k}$ is iid, according to (5.41), each channel is equally likely chosen. That is, the channel selection according to (5.41) results in the equal channel selection probability (CSP), i.e. $q_l = \frac{1}{L}$ for all l,

It is noteworthy that the equal CSP is optimal when SDs are uniformly distributed over L channels where $S_l = S$ for all l. However, if the number of SDs varies from one channel to another, the approach in (5.41) may lead to a degraded throughput as the equal CSP is no longer optimal. Thus, let us consider a channel selection approach that takes into account both the numbers of SDs, $\{S_l\}$, and channel gains, $\{G_{l;k}\}$, over L channels.

Suppose that the optimal CSPs, $\{q_l^*\}$, are available at DDs. Since optimal CSP depends on $\{\omega_l\}$, they only depend on the numbers of SDs. Thus, the BS can broadcast $\{q_l^*\}$ so that all DDs know them. For convenience, we only consider one DD.

Thus, the index of DD is omitted, and $G_l(t)$ represents the channel power gain of the DD of interest at time t.

Suppose that a DD has state variables for all L channels, which are denoted by $\{Q_l(t)\}$ at time t. The state variables are updated as follows:

$$Q_l(t+1) = \begin{cases} Q_l(t) + c(q_l^* - X_l(t)), & \text{if DD is active,} \\ Q_l(t), & \text{otherwise,} \end{cases} \tag{5.42}$$

where $c > 0$ and

$$X_l(t) = \mathbb{1}(\text{DD chooses channel } l).$$

For convenience, we assume that $c = 1$ throughout this section. Suppose that c is a sufficiently small and the DD is always active. Then, we have

$$\frac{Q_l(t)}{t} \propto q_l^* - \frac{1}{t} \int_0^t X_l(\tau) d\tau, \tag{5.43}$$

where $\frac{1}{t}\int_0^t X_l(\tau)d\tau$ can be seen as the time fraction that channel l is selected. Thus, if $Q_l(t)$ increases, channel l is underutilized. On the other hand, if $Q_l(t)$ decreases, channel l is over-utilized.

Define the system preference indicator as

$$\eta_l(t) = \exp\left(-\frac{Q_l(t)}{A(t)}\right), \quad l = 1, \ldots, L, \tag{5.44}$$

where $A(t)$ denotes the number of the slots that the DD becomes active over t slots. Here, $\frac{Q_l(t)}{A(t)}$ represents the deviation between q_l^* and the actual time fraction that channel l is used. From a system point of view, it is desirable that an active DD chooses the channel that minimizes $\eta_l(t)$ to send a packet through the most under-utilized channel. On the other hand, the user preference indicator can be defined as $G_l(t)$ that is to be maximized to minimize the transmit power. Thus, in order to take into account both the system and user preference indicators, we can consider the following channel selection criterion:

$$l^* = \arg\max_l \frac{[G_l(t)]^\alpha}{[\eta_l(t)]^\beta}, \tag{5.45}$$

where $\alpha \geq 0$ and $\beta \geq 0$ are design parameters. Note that if $\alpha = 0$ or $\beta = 0$, the user or system preference is not taken into account, respectively. As a result, if $\beta = 0$ and $\alpha > 0$, the CSP becomes $\frac{1}{L}$.

We present simulation results of the NOMA-based approach under the setting that is similar to that in Bonnefoi et al. (2018). In particular, we assume that $L = 10$ and the SDs are distributed over $L = 10$ channels as follows:

$$(S_1, \ldots, S_L) = (0.3, 0.2, 0.1, 0.1, 0.05, 0.05, 0.02, 0.08, 0.01, 0.09) \times M_1,$$

where the total number of SDs is set to $M_1 = 1000$. The total number of DDs, M_2, is set to 100 for all simulations. For NOMA, we assume that $\Gamma = \Gamma_1$. Furthermore,

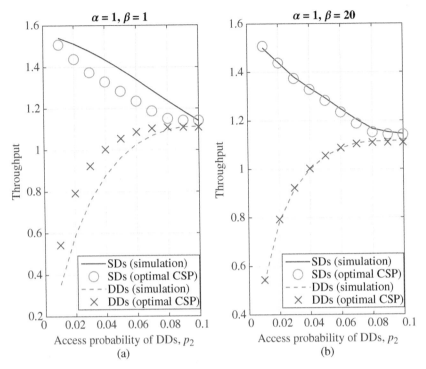

Figure 5.14 Performance of the NOMA-based approach in terms of throughput for various values of p_2 when $p_1 = 0.04$: (a) throughput when $(\alpha, \beta) = (1, 1)$; (b) throughput when $(\alpha, \beta) = (1, 20)$.

we assume that the $G_{l;k}$'s are iid and $G_{l;k} \sim \text{Exp}(1)$, which means that the channels are modeled as independent Rayleigh fading channels. For the channel selection at DDs, (5.45) is employed.

Figure 5.14 shows the performance of the NOMA-based approach in terms of the throughput when the access probability of DDs, p_2, varies. It is shown that the throughput of DDs increases with p_2, while that of SDs decreases with p_2. With a large value of β, the channel gains do not affect the channel selection. As a result, in Figure 5.14b, each DD can have the CSPs that are close to the optimal CSPs. However, a large β results in a high transmit power as shown in Figure 5.15. In Figure 5.15, the normalized transmit power under full channel-aware selection is the case that $\beta = 0$ and $\alpha > 0$. Thus, the normalized transmit power is given by $\mathbb{E}\left[min_l \frac{1}{G_l}\right]$.

The impact of α on the performance is shown in Figure 5.16 when $\beta = 1$ and $(p_1, p_2) = (0.05, 0.05)$. As α increases, the channel selection in (5.45) tends to

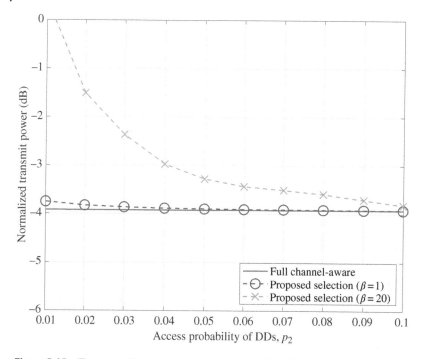

Figure 5.15 The normalized transmit power in dB of the NOMA-based approach in terms of throughput for various values of p_2 when $p_1 = 0.04$.

consider the user preference more than the system preference. Thus, the transmit power decreases with α, while the throughput of DDs decreases.

As shown in Figures 5.14 and 5.16, using the channel selection approach in (5.45) with proper choices of the values of α and β, we can maximize the throughput of DDs and/or minimize the transmit power for the NOMA-based approach.

5.4 NOMA-Based Random Access with Multiuser Detection

When applying NOMA to multichannel ALOHA in Sections 5.2 and 5.3, we assumed that the subchannels are orthogonal. As a result, no multiuser detection was considered and a simple collision model was used to find the throughput. As in Section 4.6, it is possible to consider multiuser detection when spreading codes are used to form multiple subchannels. In this section, we discuss the application of NOMA to random access with multiuser detection.

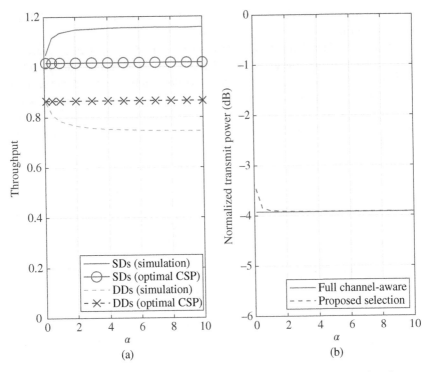

Figure 5.16 Performance of the NOMA-based approach for various values of α when $\beta = 1$ and $(p_1, p_2) = (0.05, 0.05)$: (a) throughput; (b) normalized transmit power in dB.

5.4.1 Compressive Random Access

In this section, we briefly review the model for random access with multiuser detection studied in Section 4.6.

Suppose that there are M users and each user has a unique signature sequence. Recall that $\mathbf{g}_m$ denotes the mth signature sequence or vector of length $N = N_{\text{pg}}$ and define $\mathbf{G} = [\mathbf{g}_1 \cdots \mathbf{g}_M] \in \mathbb{C}^{N \times M}$, where M is the number of signature vectors and $N < M$. At the receiver, the received signal is given by

$$\mathbf{y} = \sum_{m=1}^{M} \mathbf{g}_m s_m + \mathbf{n}$$

$$= \mathbf{G}\mathbf{s} + \mathbf{n}, \tag{5.46}$$

where $\mathbf{s} = [s_1 \cdots s_M]^{\mathsf{T}}$ represents the signal vector and $\mathbf{n}$ is the background noise vector. Here, $s_m = x_m b_m$ denotes the mth element of $\mathbf{s}$, where x_m and $b_m \in \{0, 1\}$ represent the data symbol of user m and the activity variable, respectively. We have $b_m = 1$ if user m is active. Otherwise, it becomes zero. Due to sparse activity, $\mathbf{s}$

becomes a sparse vector (i.e. only a fraction of the elements are nonzero, while the others are zero).

In compressive random access (CRA), we can consider two steps at a receiver or BS. In the first step, the active user detection (AUD) is carried out by exploiting the sparsity of $\mathbf{s}$. Here, the sparsity of a vector can be measured by the number of nonzero elements of the vector, i.e. $||\mathbf{x}||_0 = \sum_{l:\ x_l \neq 0} 1$ for a vector $\mathbf{x}$. Once the active users are identified, in the second step, the multiuser detection is performed so that the receiver is able to detect the transmitted data symbols from the active users.

5.4.2 Layered CRA

As mentioned earlier, in CRA, each user has a unique signature sequence. To support a number of users with limited bandwidth, non-orthogonal sequences can be used for signature sequences. However, since the cross-correlation increases with the number of sequences, the number of sequences cannot be arbitrarily large for a reasonable performance of AUD. In this section, in order to increase the number of sequences (and hence the number of users in CRA) without increasing the length of sequences, we apply the notion of power-domain NOMA.

Suppose that there are Q different layers in the power domain. Each layer is characterized by a different received signal power at the BS. In each layer, we assume that there are M users with unique signature sequences and the resulting multiple access channels can be characterized in two-dimensional domain (of code and power) as shown in Figure 5.17. Throughout this section, we assume that there are a total of $K = QM$ users (i.e. in each layer, there are $M = \frac{K}{Q}$ users) and the signature sequences are Gaussian.

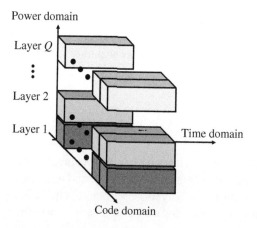

Figure 5.17 An illustration of layered CRA based on power-domain NOMA.

An active user, say user k, in the qth layer determines its transmit power such that its receive power level can be

$$P_k|h_k|^2 = v_q,$$

where v_q is the (predetermined) received signal power for layer q. Note that if P_k is greater than the maximum transmit power due to deep fading, the user cannot be active (and needs to wait till the channel fading coefficient is sufficiently large). For convenience, let $v_1 > \cdots > v_Q \,(> 0)$. In addition, let $\mathbf{G}_q = [\mathbf{g}_{q,1} \cdots \mathbf{g}_{q,M}]$, where $\mathbf{g}_{q,m}$ represents the Gaussian signature sequence for the mth user in layer q. Then, the received signal is expressed as

$$\mathbf{y}(t) = \sum_{q=1}^{Q} \mathbf{G}_q \mathbf{s}_q(t) + \mathbf{n}(t), \tag{5.47}$$

where $[\mathbf{s}_q(t)]_m = h_{q,m}\sqrt{P_{q,m}}x_{q,m}(t)b_{q,m}$. Here, $h_{q,m}$, $P_{q,m}$, $x_{q,m}(t)$, and $b_{q,m} \in \{0,1\}$ are the channel coefficient, transmit power, data symbol, and activity variable of the mth user in layer q, respectively. The resulting scheme is referred to as layered compressive random access (L-CRA).

In order to detect signals in the qth layer, let

$$\mathbf{y}_q(t) = \mathbf{y}(t) - \sum_{l=1}^{q-1} \mathbf{G}_l \mathbf{s}_l(t) = \mathbf{G}_q \mathbf{s}_q(t) + \mathbf{u}_q(t), \tag{5.48}$$

where

$$\mathbf{u}_q(t) = \sum_{l=q+1}^{Q} \mathbf{G}_l \mathbf{s}_l(t) + \mathbf{n}(t).$$

Since $\mathbb{E}[x_{q,m}(t)] = 0$, we have $\mathbb{E}[\mathbf{u}_q(t)] = 0$. Let $\mathbb{E}[\mathbf{u}_q(t)\mathbf{u}_q(t)^{\mathrm{H}}] = \sigma_q^2 \mathbf{I}$. We discuss later σ_q^2, i.e. the variance of the interference-plus-noise vector at layer q. For convenience, define the SNR at layer q as

$$\mathrm{SNR}_q = \frac{v_q}{\sigma_q^2}.$$

In addition, let K_q and ρ_q be the number of active users in layer q and the probability that a user in layer q becomes active, i.e. the access probability, respectively. Clearly, $\mathbb{E}[K_q] = M\rho_q$ as there are M users in each layer. More importantly, we can see that σ_q^2 depends on $v_{q+1}, \ldots, v_Q$ as well as $\rho_{q+1}, \ldots, \rho_Q$.

The interference vector in (5.48), i.e. $\mathbf{u}_q(t)$, is Gaussian if $q = Q$ since $\mathbf{u}_Q(t)$ only includes the background noise. For $q < Q$, since Gaussian spreading is considered (i.e. the signature vectors, $\{\mathbf{g}_{q,m}\}$, are Gaussian), $\mathbf{u}_q(t)$ can be seen as a superposition of Gaussian vectors. Thus, we can consider the following assumption[1]: $\mathbf{u}_q(t)$

---

1 In fact, the assumption in (5.49) is an approximation, because the number of Gaussian signature vectors in each layer, i.e. K_q, is random. In other words, the sum of a random number of independent Gaussian random variables is not Gaussian (but Bernoulli Gaussian).

is a Gaussian random vector, i.e.

$$\mathbf{u}_q(t) \sim \mathcal{CN}(0, \sigma_q^2 \mathbf{I}), \tag{5.49}$$

where

$$\sigma_q^2 = \sum_{l=q+1}^{Q} v_l M \rho_l + N_0. \tag{5.50}$$

In (5.50), $v_q M \rho_q$ is the variance of the signals in layer q.

As shown in (5.48), the signal detection in L-CRA can be carried out from layer 1 to layer Q. At layer q, taking $\mathbf{u}_q(t)$ as the background noise, the AUD can be considered with M users and $\mathbf{G} = \mathbf{G}_q$. As a result, up to $M - 1$ active users can be detected in each layer provided that SNR_q is sufficiently high.

In L-CRA, since there always exists the interference due to the signals in higher layers, the success of SIC depends on the background noise and interference (which also depends on the number of layers and power allocation). In general, for successful SIC in all layers with a high probability, it is necessary to carefully decide the power levels, $\{v_q\}$.

As shown in Figure 5.18, suppose that a cell is divided into Q regions. Region q is a circular ring with the inner and outer radii, R_{q-1} and R_q, respectively, where $R_0 = 0$. In this case, since the transmit power of a user in region q needs to be higher than that in region $q - 1$ for the same receive power, the users in region q can be assigned to layer q in order to avoid a high transmit power (Choi, 2017b).

If K users are assumed to be uniformly distributed in a cell. Then, $\{R_q\}$ can be decided to make the area of each region equal (so that each area has the same number of users (on average), $M = \frac{K}{Q}$). For normalization purpose, we assume that $R_Q = 1$ (i.e. the cell size is normalized). Since the power control is considered for the predetermined receive power levels, for a user in region or layer q, we expect

$$P_{q,m} \propto R_q^\kappa v_q,$$

where κ is the path loss exponent. Here, we consider a user on the outer ring (i.e. the worst case) and $|h_k|^2 \propto \frac{1}{R_q^\kappa}$.

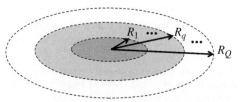

Figure 5.18 An illustration of dividing a cell into Q regions to reduce the transmit power.

According to Choi (2019b), with a target SNR, denoted by Γ, for a given access probability $\rho = \rho_q$ for all q, the power levels can be recursively decided as follows:

$$v_q = \frac{\Gamma}{\beta\left(\frac{1}{\sigma_q^2}, \zeta\right)}, \quad q = Q, \dots, 1, \tag{5.51}$$

where $\zeta = \frac{\rho M}{N}$ and

$$\beta(\gamma, \zeta) = \frac{(1-\zeta)\gamma}{2} - \frac{1}{2} + \sqrt{\frac{(1-\zeta)^2\gamma^2}{4} + \frac{(1+\zeta)\gamma}{2} + \frac{1}{4}}. \tag{5.52}$$

Here, for $q = Q$, since $v_Q = \frac{\Gamma}{\beta\left(\frac{1}{N_0}, \zeta\right)}$, we need to have

$$\Gamma \geq \beta\left(\frac{1}{N_0}, \zeta\right) \tag{5.53}$$

to make sure that $v_Q \geq 1$. Note that since $\beta(\gamma, \zeta)$ increases with γ and $\sigma_q^2 > \sigma_{q+1}^2$, from (5.51), we can see that $v_1 > \cdots > v_Q$ as expected.

For example, suppose that $Q = 3$, $M = 100$, $N = 30$, and $\rho = \rho_q = 0.05$ for all q. Then, there might be 15 active users on average. In addition, let $\Gamma = 4$, $\kappa = 3.5$, and $N_0 = 1$. In Table 5.1, the predetermined power levels, v_q's, and the average transmit powers for three layers are shown, where we can see that by taking advantage of shorter distances to the BS, the transmit power of the users in region q can be lower than v_q for $q < Q$.

Note that the average transmit power of the users in layer 1 might be too high although the advantage of the short distance to the BS in layer 1 is taken into account when $Q = 3$ as shown in Table 5.1. Thus, the number Q of layers should not be too large to avoid high transmit power (e.g. L-CRA with two layers would be preferable in practice).

5.4.3 Performance Under Realistic Conditions

In order to analyze the performance of CRA in Section 4.6.2, the notion of multiple measurement vector (MMV) is considered, i.e. it has been assumed that the

Table 5.1 The transmit power for each region and the predetermined power level for each layer.

Layer	R_q	v_q (dB)	Transmit power (dB)
1	0.577	32.840	24.490
2	0.816	19.618	16.536
3	1	6.382	6.382

receiver is able to recover a sparse signal if the number of active devices is less than or equal to $N - 1$. This is valid under various assumptions including a high SNR. However, in practice, due to fading, interference, and noise, the receiver (i.e. BS) may not be able to correctly detect a sparse signal or the signature sequences of active devices, which results in the following two possible errors:

- false alarm (FA) error: If the receiver detects a signature sequence that is not transmitted;
- missed detection (MD) error: If the receiver fails to detect a signature sequence that is transmitted by an active user.

In CRA, due to FA errors, the BS can send acknowledgment (ACK) to inactive users and those inactive users can ignore the feedback signals. On the other hand, the active users associated with MD errors do not receive any feedback signals. In this case, those active users can assume that their signals are not successfully transmitted and perform retransmissions. As a result, it can be seen that MD errors have a more serious impact on performance than FA errors.

However, in L-CRA, both FA and MD errors result in error propagation. That is, any FA and MD errors in layers 1 to $q - 1$ can create interference in layer q as shown in (5.48). Thus, the number of layers, Q, cannot be large to avoid not only high transmit power as mentioned earlier but also severe error propagation.

In Figure 5.19, we show the average number of FA/MD errors when $Q = 3$ and $K = 300$ (thus, $M = 100$ users per layer) as functions of access probability ρ. Here, it is assumed that the access probability is the same for all layers and the BS knows the number of the active users. Thus, the number of FA errors is equal to that of FA errors. For convenience, users with FA and MD errors are referred to as FA and MD users, respectively. It is clearly shown that error propagation occurs as the average number of MD/FA users increases with the index of layer, q, and also increases with the access probability, ρ. Note that when $\rho = 0.1$, the average number of active users per layer is 10. Thus, when $Q = 3$, according to Figure 5.19, most active users in layer 3 are incorrectly detected (i.e. about 8 out of 10). On the other hand, if $\rho = 0.05$, the average number of MD/FA users is approximately 1 in layer 3, which means that approximately 20% of the active users in layer 3 are incorrectly detected. This implies that the access probability should be low enough to avoid excessive retransmissions.

5.5 Further Reading

In ALOHA, the signal power difference of the received signals has already been exploited as capture effect, while NOMA-ALOHA paves a way to systematically

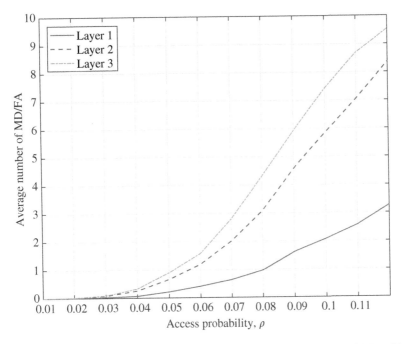

Figure 5.19 Average numbers of FA/MD errors when $Q = 3$ and $K = 300$ (thus, $M = 100$ users per layer) as functions of access probability ρ.

exploit capture effect in ALOHA (Choi, 2017b). The reader is referred to Roberts (1975), Metzner (1976), and Namislo (1984) to see early works on capture effect.

Although power-domain NOMA is not applied, it is still possible to employ SIC to improve the throughput of ALOHA with (coded) repetition diversity as shown in Casini et al. (2007) and Liva (2011). SIC can also be used in interactive collision resolution (Yu and Giannakis, 2007) and retransmission diversity based multiple access (Choi, 2016).

6

Application of NOMA to MTC in 5G

In this chapter, we explain existing approaches employed for machine-type communication (MTC) that will play a key role in supporting the connectivity of sensors and devices for a number of Internet-of-Things (IoT) applications such as smart cities, factory automation, and so on. We then discuss how non-orthogonal multiple access (NOMA) can be used to improve the performance of MTC and other IoT networks.

6.1 Machine-Type Communication

6.1.1 IoT Connectivity

The IoT is a network of things (i.e. smart phones, smart meters, and devices including sensors and actuators, etc.) that are connected through the Internet for a number of applications (Gubbi et al., 2013; Kim et al., 2016), which include smart home, wearables, smart cities, smart grids, smart factories, connected autonomous vehicles, and so on. To support the connectivity, a number of different approaches have been considered. For example, well-known ZigBee, Bluetooth, and WiFi can be used to connect devices for short-range connectivity. While small-scale IoT systems such as smart home can rely on short-range connectivity, large-scale IoT systems (e.g. environmental monitoring systems with sensors distributed over a city or suburbs) should require long-range connectivity. To support long-range connectivity with low power consumption in unlicensed bands, low-power wide area networks (LPWANs) has been proposed. In addition, MTC; (3GPP, 2016; 2018) can provide long-range connectivity for devices and sensors over a large geographical area (regions or countries) for cellular IoT applications. It is noteworthy that MTC is part of cellular systems and uses licensed bands, while LPWAN uses unlicensed band. Thus, a private IoT network can be built using LPWAN. On the other hand,

Massive Connectivity: Non-Orthogonal Multiple Access to High Performance Random Access,
First Edition. Jinho Choi.
© 2022 The Institute of Electrical and Electronics Engineers, Inc. Published 2022 by John Wiley & Sons, Inc.

MTC can be used to support IoT applications that need extensive coverage (e.g. a region, state, or nation) through cellular networks.

The notion of massive MTC is devised to support a large number of devices in cellular IoT. For example, a minimum connection density of 1 million devices for every square kilometer is expected in massive MTC. An overview of massive MTC can be found in Bockelmann et al. (2016) and a survey for various approaches to IoT connectivity solutions can be found in Ding et al. (2020a).

6.1.2 Random Access Schemes for MTC

In MTC, due to sparse activity and sporadic traffic of devices and sensors, random access based schemes are employed to transmit the information of devices, because their signaling overhead is low. Most random access schemes use a set of preambles that are sequences of zero or low cross-correlation. In principle, a random access system can be seen as a multiuser system, and the use of a set of preambles can allow simultaneous detection of multiple active devices and their channel estimation. One of well-known random access schemes for MTC is a four-step random access scheme (3GPP, 2018), which is illustrated in Figure 6.1. The first step is random access to establish connection to the base station (BS) with a pool of preambles consisting of L sequences. In the first step, an active user equipment (UE) transmits a randomly selected preamble through physical random access channel (PRACH). For convenience, UE and device are used interchangeably in this chapter. In the second step, the BS detects the transmitted preambles and sends responses, in which resource blocks (RBs) or channels,[1] called physical uplink shared channel (PUSCH), allocated to the UEs are specified. Then, UEs can transmit data packets with their identification sequence in the third step through the allocated PUSCH. Notice that if there are multiple UEs that choose the same preamble at the first step, they transmit their signals on the same PUSCH, which leads to a collision and decoding failure. The resulting collision is referred to as a preamble collision. If there are no preamble collisions, the data packets from UEs at the third step can be successfully decoded. Then, in the fourth step, each UE receiving the downlink message will compare the identity sequence in the message with the identity sequence transmitted in the third step. If a UE finds that the received identity sequence in the fourth step is the same as that transmitted as part of the third step, it can declare successful transmission. Otherwise, the UE deems that its signal collides with others and may attempt retransmission after a random amount of time. As a result, a UE can know whether or not the transmission was successful only after the fourth step is completed.

---

1 We will use the terms resource block and channel interchangeably.

Figure 6.1 An illustration of four-step random access in MTC.

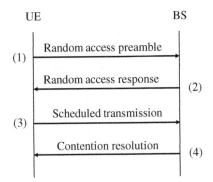

Figure 6.2 An illustration of two-step random access in MTC.

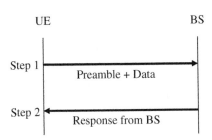

To further reduce signaling overhead, two-step random access (3GPP, 2018) rather than four-step random access can be considered if all devices have short messages to transmit. In Figure 6.2, an illustration of two-step random access is shown. The first and third steps of four-step random access are combined into the first step of two-step random access. That is, a UE transmits a data packet without waiting for feedback after transmitting a randomly selected preamble in the first step. In the second step, the BS is to send a feedback signal so that devices can see whether or not transmissions are successful.

The first step of two-step random access consists of two phases: the preamble and data transmission phases. A randomly selected preamble is transmitted in the first phase and then data packet transmission in the second phase as shown in Figure 6.3. Since a data packet is transmitted with a preamble in one step (i.e. the first step) without waiting for the response from the BS (which is the second step of conventional four-step random access), two-step random access is also called grant-free random access.

Figure 6.3 Two phases (i.e. preamble transmission and data transmission phases) for the first step in two-step random access.

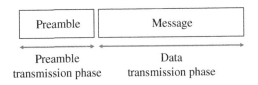

In two-step random access, there must be orthogonal subchannels in the PUSCH for uplink transmission, and each subchannel must correspond to one of the preambles. That is, if an active UE chooses the *l*th preamble in the preamble transmission phase, it needs to transmit its packet through the *l*th subchannel in the data transmission phase. Thus, if an active UE experiences a preamble collision, its packet will be collided. From this, we can see that it is possible to carry out the performance analysis of two-step random access using the model of multichannel ALOHA (discussed in Chapter 4), where the number of multichannels is equal to that of preambles.

6.2 A Model with Massive MIMO

In this section, we consider the case that the BS is equipped with a large number of antennas so that the spatial selectivity can be exploited in multiple access. In two-step random access, this means that a single RB can be shared by multiple active UEs for PUSCH. For convenience, in this chapter, the terms UE and device are used interchangeably.

6.2.1 Massive MIMO

In massive multiple input multiple output (MIMO), due to a large number of service antennas co-located at a BS, the channel hardening (i.e. the effect of small-scale fading is averaged out and devices' channels behave deterministic like a wired channel) and favorable propagation (i.e. the propagation channels to different devices become orthogonal, which makes different devices distinguishable in the space domain) can be exploited.

Suppose that the BS is equipped with M antennas and each device one antenna. Let $\mathbf{h}_k \in \mathbb{C}^M$ denote the channel vector from device k to the BS. The mth element of $\mathbf{h}_k$, denoted by $h_{m,k}$, is the channel coefficient from device k to the mth antenna at the BS, which is given by

$$h_{m,k} = v_{m,k} \sqrt{\alpha_k}, \qquad (6.1)$$

where α_k represents the large-scale fading term and $v_{m,k} \sim \mathcal{CN}(0,1)$ is the small-scale fading term for Rayleigh fading. In general, α_k depends on the distance between device k and the BS. In addition, since all the antennas are located at the BS, the large-scale fading term, α_k, is applied to all the channel coefficients as in (6.1), i.e. $\mathbf{h}_k = \sqrt{\alpha_k}\mathbf{v}_k$, where $\mathbf{v}_k = [v_{1,k} \cdots v_{M,k}]^{\mathrm{T}}$. Since there are a number of devices for multiple inputs and a number of antennas at the BS for multiple outputs, the resulting system is a MIMO system. The notion of massive MIMO is related to the law of large numbers for a large M.

Suppose that the signal received through a large antenna array is to be combined with a weighting vector $\mathbf{w} = [w_1 \cdots w_M]^{\mathrm{T}}$. The signal combining operation with the outputs of multiple antennas is called beamforming. The output after beamforming is given by

$$r = \sum_{m=1}^{M} w_m y_m, \tag{6.2}$$

where y_m represents the received signal at the mth antenna. Assume that device k is active and the other devices are inactive. Then, the received signal becomes

$$y_m = h_{m,k} \sqrt{P_k} s_k + n_m, \quad m = 1, \dots, M, \tag{6.3}$$

where P_k and s_k are the transmit power and transmitted signal of device k, respectively, and $n_m \sim \mathcal{CN}(0, N_0)$ is the background noise at antenna m. For normalization, we assume that $\mathbb{E}[|s_k|^2] = 1$. Then, we have

$$
\begin{aligned}
r &= \sum_{m=1}^{M} w_m y_m \\
&= \sum_{m=1}^{M} w_m h_{m,k} \sqrt{P_k} s_k + \sum_{m=1}^{M} w_m n_m,
\end{aligned} \tag{6.4}
$$

where the first term on the RHS is the signal and the second term is the noise. The SNR becomes

$$
\begin{aligned}
\mathrm{SNR} &= \frac{P_k |\sum_{m=1}^{M} w_m h_{m,k}|^2}{\mathbb{E}[|\sum_{m=1}^{M} w_m n_m|^2]} \\
&= \frac{P_k |\sum_{m=1}^{M} w_m h_{m,k}|^2}{N_0 \sum_{m=1}^{M} |w_m|^2} \\
&\leq \frac{P_k \sum_{m=1}^{M} |w_m|^2 \sum_{m=1}^{M} |h_{m,k}|^2}{N_0 \sum_{m=1}^{M} |w_m|^2} \\
&= \frac{P_k \sum_{m=1}^{M} |h_{m,k}|^2}{N_0} \\
&= \frac{P_k \alpha_k \sum_{m=1}^{M} |v_{m,k}|^2}{N_0},
\end{aligned} \tag{6.5}
$$

where the inequality is due to the Cauchy–Schwarz inequality. The equality can be achieved when $w_m \propto h_{m,k}^*$. If the transmit power, P_k, is decided to compensate the large-scale fading term, α_k, and inversely proportional to M, the SNR in (6.5) is given by

$$\mathrm{SNR} = C \frac{1}{M} \sum_{m=1}^{M} |v_{m,k}|^2, \tag{6.6}$$

where $C > 0$ is a constant. Since each $v_{m,k}$ is independent, due to the law of large numbers, we have

$$\lim_{M \to \infty} \text{SNR} = C\mathbb{E}[|v_{m,k}|^2] = C \tag{6.7}$$

with probability 1, which shows the channel hardening phenomenon. That is, the SNR in (6.5), which was a random variable with a small M, becomes a constant as M increases.

For two different devices, say devices 1 and 2, we can show that

$$\frac{1}{M} \mathbf{h}_1^H \mathbf{h}_2 = \sqrt{\alpha_1 \alpha_2} \frac{1}{M} \sum_{m=1}^{M} v_{1,m} v_{2,m}^* \to 0, \quad M \to \infty. \tag{6.8}$$

Thus, the channel vectors are asymptotically orthogonal as $M \to \infty$, and the resulting phenomenon is referred to as favorable propagation in the context of massive MIMO.

When there are K devices transmitting their signals simultaneously, the BS receives the following signal:

$$\mathbf{y} = \mathbf{h}_1 \sqrt{P_1} s_1 + \cdots + \mathbf{h}_K \sqrt{P_K} s_K + \mathbf{n}. \tag{6.9}$$

Due to the channel hardening and favorable propagation of massive MIMO, if the BS wants to see the signal from device k, it only needs to perform the following operation:

$$\frac{1}{M} \mathbf{h}_k^H \mathbf{y} = \frac{1}{M} \mathbf{h}_k^H \mathbf{h}_1 \sqrt{P_1} s_1 + \cdots + \frac{1}{M} \mathbf{h}_k^H \mathbf{h}_K \sqrt{P_K} s_K + \frac{1}{M} \mathbf{h}_k^H \mathbf{n}$$
$$\approx C s_k, \quad \text{for a large } M. \tag{6.10}$$

Based on (6.10), massive MIMO seems to allow an infinite number of coexisting signals, K, i.e. $K \to \infty$, as $M \to \infty$ (Björnson et al., 2018).

As shown in (6.10), a channel vector can be seen as a key to open a drawer of a multi-drawer cabinet as illustrated in Figure 6.4. The number of drawers can approach infinite as $M \to \infty$, which means that a large number of signals can coexist, while any of them can be easily extracted as in (6.10). However, in order to see any of them, the receiver has to have a key to the corresponding signal, which is the channel vector. Therefore, in massive MIMO, it is important to estimate the channel vectors of all the active devices. To this end, each device can have a unique pilot signal that allows the BS to estimate its channel vector. Furthermore, the pilot signals should be orthogonal to each other. Otherwise, the estimated channel vector may have interfering signals due to other non-orthogonal pilot signals. In this case, the BS may not be able to recover the signal transmitted by a device of interest, because the key, i.e. the estimated channel vector, is not reliable.

The need of unique orthogonal pilot signals results in a dilemma, because the length of orthogonal pilot signals grows linearly with the number of devices. That

Figure 6.4 A multi-drawer cabinet as a received signal in a massive MIMO system, where the channel vectors are the keys to open drawers.

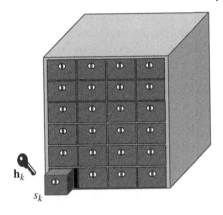

is, massive MIMO allows a large number of devices to coexist without interfering with each other, but requires very long pilot signals to estimate the channel vectors required for successful signal decoding, which offsets the advantage of massive MIMO.

6.2.2 Two-Step Random Access with Massive MIMO

In two-step random access for MTC, the two features of massive MIMO, namely the channel hardening and favorable propagation, can be exploited to allow multiple MTC devices to transmit their data packets through the shared channel RB in the PUSCH. From this, with a limited bandwidth, a large number of devices can be supported.

As mentioned earlier, since the BS needs to estimate the channel vectors of active devices, in two-step random access, an active device is to choose one preamble from a pool of preambles to transmit it as a pilot signal (as a result, each device does not need to have a unique long pilot signal at the cost of preamble collision that will be discussed later). Furthermore, massive MIMO can make two-step random access more efficient by allowing transmissions of preambles and data on the same channel. In particular, for the first step that has both the preamble and data transmission phases, one time slot can be used. Each active device chooses a preamble from a pool of L preambles uniformly at random, and transmits it in the preamble transmission phase. In the data transmission phase, a data packet is then transmitted. Since all the devices are synchronized in MTC, it is expected that the length of data packet is the same for all devices (or the length of data transmission phase is decided by the maximum length of data packet among all the devices).

In summary, with massive MIMO, two-step random access becomes more efficient as follows: (i) a single RB can be used for transmitting both preambles and data packets and (ii) thanks to the orthogonality among those spatial channels, all the active devices can transmit their data packets through the same shared channel during the data transmission phase without any orthogonal multiple access schemes (e.g. time division multiple access).

6.2.3 Throughput Analysis

To decode coexisting signals via beamforming in massive MIMO, the BS needs to estimate the spatial channels of active devices. An active device's channel can be estimated if its preamble is not transmitted by other active devices (i.e. there is no preamble collision). Once the BS succeeds to estimate the channel vector of an active device, it can decode the data packet transmitted by this active device during the data transmission phase. Consequently, the number of the active devices that succeed to transmit their data packets in two-step random access is equal to the number of the active devices that do not experience preamble collisions.

Suppose that there are K active devices and L preambles. Then, the average number of the active devices that do not experience preamble collisions is given by

$$\eta(K) = K \Pr(\text{no collision of an active device} \mid K)$$
$$= K \left(1 - \frac{1}{L}\right)^{K-1}, \tag{6.11}$$

because the probability that the preamble chosen by a particular active device is not chosen by the other $K - 1$ active devices is $\left(1 - \frac{1}{L}\right)^{K-1}$, which is identical to (4.5). Thus, the throughput analysis of two-step random access can be carried out using the model of multichannel ALOHA in Section 4.2.3. That is, if K follows a Poisson distribution with mean λ, the throughput becomes

$$\eta_{2S} = \mathbb{E}\left[K\left(1 - \frac{1}{L}\right)^{K-1}\right]$$
$$= \lambda e^{-\frac{\lambda}{L}} \ (\leq L e^{-1}), \tag{6.12}$$

since the moment generating function (MGF) of Poisson distribution, i.e. $M_K(z) = \mathbb{E}[z^K] = e^{\lambda(z-1)}$. Here, λ is the average number of active devices per slot.

In the throughput expression in (6.12), there does not seem to be any benefit of massive MIMO as the throughput is given as the average number of the active devices that succeed to transmit their packets *per slot*. To see the gain from massive MIMO, we need to consider the two systems, one without massive MIMO and the other with massive MIMO, and find the normalized throughput in unit time. As mentioned earlier, the slot consists of the two phases. We assume that the duration of the preamble transmission phase is identical for the two cases. Furthermore, we assume that the length of preamble transmission phase is the number of preambles, L, in unit time. The length of the data transmission phase without massive MIMO is proportional to the number of preambles and given by LD, where D is the length of data packet in unit time. Then, the length of slot without massive MIMO becomes

$$
\begin{aligned}
T_{\text{wo}} &= L + LD \\
&= L(1 + D) \quad \text{(in unit time)}.
\end{aligned}
$$

(6.13)

On the other hand, as mentioned earlier, with massive MIMO, all active devices transmit their data packets through the shared channel. Thus, the length of the data transmission phase becomes D, and the length of slot becomes

$$
\begin{aligned}
T_{\text{w}} &= L + D \\
&= L + D \quad \text{(in unit time)}.
\end{aligned}
$$

(6.14)

For convenience, denote by ρ the average number of active devices per unit time. Then, without massive MIMO, the average number of active devices per slot becomes $\lambda_{\text{wo}} = \rho T_{\text{wo}} = \rho L(1 + D)$. With massive MIMO, we have $\lambda_{\text{w}} = \rho T_{\text{w}} = \rho(L + D)$.

To compare the performance of two-step random access with/without massive MIMO, we consider the following normalized throughput with/without massive MIMO (in the number of bits per unit time):

$$
\begin{aligned}
\overline{\eta}_{2S,\text{wo}} &= \frac{\eta_{2S,\text{wo}} RD}{T_{\text{wo}}} = \rho RD e^{-\rho(1+D)}, \\
\overline{\eta}_{2S,\text{w}} &= \frac{\eta_{2S,\text{w}} RD}{T_{\text{w}}} = \rho RD e^{-\rho\left(1+\frac{D}{L}\right)},
\end{aligned}
$$

(6.15)

where R represents the number of bits per unit time. Note that if the unit time can be the duration of data symbol in modulation, R becomes the number of bits per data symbol. From (6.15), the throughput ratio can be given by

$$
\kappa = \frac{\overline{\eta}_{2S,\text{w}}}{\overline{\eta}_{2S,\text{wo}}} = e^{\rho D\left(1-\frac{1}{L}\right)},
$$

(6.16)

which shows that the throughput gain due to massive MIMO increases with ρD. In other words, the higher the traffic intensity (ρ) or the longer the data packets (D), the higher the throughput gain, κ, by massive MIMO.

It is noteworthy that the normalized throughput with massive MIMO in (6.15) is bounded as follows:

$$
\begin{aligned}
\overline{\eta}_{2S,\text{w}} &= \frac{\rho DR\left(1 + \frac{D}{L}\right)}{1 + \frac{D}{L}} e^{-\rho\left(1+\frac{D}{L}\right)} \\
&\leq \frac{LDR}{L+D} e^{-1}.
\end{aligned}
$$

(6.17)

This shows that the normalized throughput can be maximized if $L = D = \frac{T_w}{2}$, and this maximum linearly increases with T_w, i.e.

$$\bar{\eta}_{2S,w} \leq \frac{LDR}{L+D}e^{-1}$$
$$\leq \frac{Re^{-1}}{4}T_w. \qquad (6.18)$$

There are few remarks as follows:

- The normalized throughput cannot be arbitrarily high by increasing T_w. If T_w exceeds the coherence time, the estimated channel cannot be used for decoding. Thus, T_w has to be finite, which limits the normalized throughput.
- As illustrated in Figure 6.4, although a cabinet (i.e. massive MIMO) can have a large number of drawers (i.e. signals from active devices), contents can only be accessed with reliable keys. Unfortunately, the cost of finding keys cannot be overlooked, which rather becomes the main difficulty. This can be confirmed by (6.18), since the maximum throughput can be achieved only when a half of resource (i.e. T_w) is allocated to preamble transmission.

6.3 NOMA for High-Throughput MTC

To fully utilize the spatial selectivity of massive MIMO in two-step random access for higher throughput, the notion of NOMA can be considered. In this section, we discuss a NOMA-based approach where preambles and data packets can coexist.

6.3.1 Coexisting Preambles and Data Packets

For a NOMA-based approach, we assume that the length of slot is equivalent to that of preamble. In addition, the length of packet is the same as that of preamble, i.e. $T_w = L = D$. If a device has more than one packet (say up to B packets), it can continuously transmit them over multiple slots. Thus, if a device has two packets to transmit, it can send a randomly selected preamble at slot t. Provided that the BS is able to detect the selected preamble without collision, which means that the BS can estimate the channel vector of the device, this device can transmit two packets over slots $t + 1$ and $t + 2$. Since the BS is able to accurately detect whether or not a preamble is collided with the assistance of massive MIMO, it can feedback this information after receiving preambles so that devices with preamble collisions can drop their packets or retransmit after random back-off time.

In Figure 6.5, the NOMA-based approach is illustrated. In this NOMA-based approach, each device's signal can be characterized by its channel vector, not transmit power. For convenience, it is referred to as the coexisting preamble and data

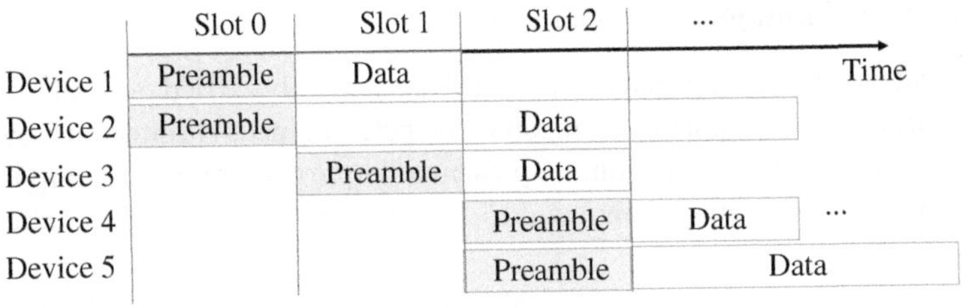

	Slot 0	Slot 1	Slot 2	...	
Device 1	Preamble	Data			Time
Device 2	Preamble		Data		
Device 3		Preamble	Data		
Device 4			Preamble	Data ...	
Device 5			Preamble	Data	

Figure 6.5 An illustration of the CoPD approach where simultaneous preamble and data transmissions are allowed.

(CoPD) transmissions approach. Let slot 0 be the initial slot where only preambles are transmitted. The BS detects the transmitted preambles. Suppose that the BS detects two preambles (transmitted by devices 1 and 2) in slot 0 and sends feedback signals to the devices. Device 1 has one data packet, which is transmitted over slot 1, and device 2 has three data packets, which are transmitted over slots 1–3. In slot 1, the BS also detects one preamble (transmitted from device 3) and sends a feedback signal at the end of slot 1. Device 3 that transmits a preamble in slot 1 has one packet that is transmitted over slot 2. The BS can detect two preambles (transmitted by devices 4 and 5) in slot 2 and sends feedback signals to the corresponding devices so that they can start to send their packets from slot 3. For convenience, the first slot that an active device transmits its preamble is referred to as the P-slot and the subsequent slots in which the active device transmit data packets are referred to as the D-slots. Thus, for device 3, slots 1 and 2 are the P- and D-slots, respectively.

We have the following assumption for the CoPD approach:

(A1) Suppose that an active device with m packets transmits a preamble in slot t. If it receives acknowledgment (ACK) at the end of slot t, it can transmit m packets from slot $t + 1$ to slot $t + m$. In addition, this device can only be active again after slot $t + m$.

In the CoPD approach, there are two different types of active devices in every slot t: (i) one group of the active devices that transmit preambles and (ii) the other group of the active devices that transmit data packets. An active device in the former group is referred to as a P-device, while an active device in the latter group a D-device. According to the assumption of A1, the two groups are exclusive.

Due to massive MIMO, the subspace spanned by the channel vectors of P-devices are asymptotically orthogonal to that spanned by the channel vectors of D-devices. In other words, the CoPD transmissions should not interfere with each other as $M \to \infty$.

6.3.2 Maximum Throughput Comparison

Under the ideal case where the channel estimation is perfectly carried out and the channel vectors are asymptotically orthogonal, the performance is limited only by preamble collisions. In this subsection, we present a simple maximum throughput comparison between the conventional and CoPD approaches under ideal conditions and a simple setting.

Suppose that the length of slot is $T = T_w$ in unit time and consider the following assumption:

(A2) The number of new active devices follows a Poisson distribution with parameter λ and is independent at each slot.[2] Thus, the average number of the new active devices per slot is λ.

Note that λ can be seen as an effective arrival rate when backlogged devices are considered. The backlogged packets are to be retransmitted after random back-off times. Thus, the effective arrival rate is given by $\lambda = \lambda_{new} + \lambda_{backoff}$, where λ_{new} and $\lambda_{backoff}$ represent the average new packet arrival and backlogged packet arrival rates, respectively. Under the assumption that backlogged packet arrivals are independent of new packet arrivals, the sum of two arrivals can be seen as a Poisson random variable with rate λ. Therefore, the assumption of A2 is still valid in this case.

In the conventional approach, we have $D = T - L$ so that an active device can transmit a preamble and data packet within a slot. Denote by A the number of the active devices without preamble collisions. Provided that there are K active devices, the conditional normalized throughput in the conventional approach is given by

$$\tilde{\eta}_{conv}(K) = \mathbb{E}[A|K]\frac{DR}{T},$$

where $\mathbb{E}[A|K]$ is the conditional mean of the number of active devices without preamble collision for given K. Since there are L preambles, the conditional mean of A is

$$\mathbb{E}[A|K] = K\left(1 - \frac{1}{L}\right)^{K-1}.$$

Therefore, we have

$$\tilde{\eta}_{conv}(K) = K\left(1 - \frac{1}{L}\right)^{K-1}\frac{DR}{T}. \tag{6.19}$$

Under the assumption of (A2), i.e. $K \sim \text{Pois}(\lambda)$, the throughput becomes

$$\tilde{\eta}_{conv} = \mathbb{E}[\tilde{\eta}_{conv}(K)]$$

---

2 In practice, each slot can consist of multiple channels over frequency and each channel can have a number of active devices. Since each channel is independent in frequency, we only focus on a single channel scenario in this section.

$$= \frac{DR}{T}\lambda e^{-\frac{\lambda}{L}}$$

$$= \frac{LDR}{T}\frac{\lambda}{L}e^{-\frac{\lambda}{L}} \leq \frac{LDR}{T}e^{-1}, \tag{6.20}$$

where the upper-bound is achieved when $\lambda = L$. Note that the upper bound in (6.20) is identical to that in (6.17). To maximize the normalized throughput, as in (6.18), L can be further optimized as follows:

$$\max \tilde{\eta}_{conv} = \max_{0 \leq L \leq T} \frac{e^{-1}}{T}(T - L)LR$$

$$= \frac{RTe^{-1}}{4}, \tag{6.21}$$

where the maximum is achieved when $L = \frac{T}{2}$. Thus, for the conventional approach, the conditions for the maximum throughput are $\lambda = L = \frac{T}{2}$.

On the other hand, in the CoPD approach, as illustrated in Figure 6.5, the length of preamble and that of data packet are the same, which is equivalent to that of slot (i.e. $L = D = T$) thanks to the design that allows coexistence of preamble and data. In addition, since all the slots can be used for data transmission in the CoPD approach, preamble overhead is virtually eliminated. Provided that there are K active devices in the CoPD approach, the conditional throughput can be similarly derived as in (6.19) and is given by

$$\tilde{\eta}_{CoPD}(K) = K\left(1 - \frac{1}{L}\right)^{K-1}\frac{TR}{T}$$

$$= K\left(1 - \frac{1}{T}\right)^{K-1}R$$

$$= RK\left(1 - \frac{1}{T}\right)^{K-1}. \tag{6.22}$$

Thus, the throughput becomes

$$\tilde{\eta}_{CoPD} = \mathbb{E}[\tilde{\eta}_{CoPD}(K)]$$

$$= R\lambda e^{-\frac{\lambda}{T}}$$

$$= RT\frac{\lambda}{T}e^{-\frac{\lambda}{T}} \leq RTe^{-1}, \tag{6.23}$$

where the upper-bound is achieved if $\lambda = T$. From (6.21) and (6.23), we can claim that the maximum throughput of the CoPD approach is four times higher than that of the conventional approach.

In Figure 6.6, the normalized throughput curves of the conventional and CoPD approaches are shown as functions of the traffic intensity, λ, when $R = 2$, $T = 30$, and $B = 1$. Then, the maximum normalized throughput of the conventional approach becomes $\frac{RTe^{-1}}{4} \approx 5.5$. On the other hand, the maximum normalized throughput of the CoPD approach is $RT\,e^{-1} \approx 22$, which can be achieved when $\lambda = T = 30$.

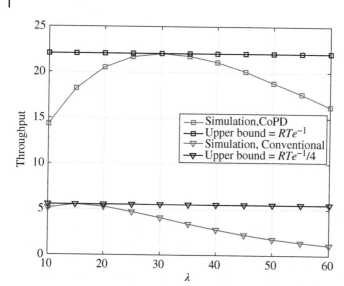

Figure 6.6 Throughput comparison between conventional and CoPD approaches with $R = 2$, $B = 1$, and $T = 30$. While the conventional approach has the maximum throughput when $L = D = \frac{T}{2} = 15$, while the CoPD approach has the maximum when $T = L = D = 30$.

Note that $\lambda = \rho T$, where ρ is the average number of active devices or (traffic intensity) per unit time, as mentioned earlier. Thus, for a fixed ρ, we have

$$\tilde{\eta}_{\text{CoPD}} = RT\rho e^{-\rho}. \tag{6.24}$$

This implies that the normalized throughput increases with T. To see how this happens, suppose that T increases. Then, a larger number of active devices can transmit signals simultaneously in a slot as $\lambda = \rho T$. As illustrated in Figure 6.4, thanks to massive MIMO (with $M \to \infty$), those signals can coexist without interfering with each other and the BS is able to decode them (with known channel vectors), which leads to an increasing throughput as T increases.

6.3.3 Limitations

In practice, unfortunately, the throughput of the CoPD approach cannot be arbitrarily high due to various limiting factors. First of all, T cannot be arbitrarily long as the coherence time is limited.

In addition, the maximum number of data packets that a device can consecutively transmit, B, cannot be large, because the coherence time is limited as mentioned earlier. Even if the coherence time can be long (for static devices), B has to be limited to avoid the error propagation. To see this, consider the example in

Figure 6.5. Provided that devices 1 and 2 transmit different preambles in slot 0, the BS can estimate the channel vectors of devices 1 and 2, and successfully to decode the data packets transmitted from them in slot 1. Then, the BS can perform successive interference cancelation (SIC) to remove the data packets from devices 1 and 2 in slot 1 so that it can estimate the channel vector of device 3, which is a P-device in slot 1. In performing SIC, due to the channel estimation error, SIC is not perfect and the residual signal remains. This results in the increase of the background noise and increases the channel estimation error for P-devices in subsequent slots. As a result, without limiting the number of D-devices in a slot, due to the increase of the channel estimation error via error propagation, the BS may fail to decode data packets. Consequently, due to limited number of D-devices and maximum number of data packets, the throughput becomes limited.

6.4   Layered Preambles for Heterogeneous Devices

In this section, we apply power-domain NOMA to the design of preambles in conventional two-step random access (where preambles and data packets do not coexist) to increase the number of preambles without bandwidth expansion.

6.4.1   Heterogeneous Devices in MTC

We assume that each device is to choose a preamble from the pool uniformly at random. This means that all the devices have the same access performance. However, devices may require different access performance. For example, a group of devices may have data sets to be transmitted subject to a low access delay requirement in real-time IoT applications. Thus, it is necessary to consider heterogeneous devices in terms of access performance requirements and the set of preambles can be divided into multiple subsets to support different access performance.

For simplicity, we only consider two different types of devices in this section as follows:

- Type-1 devices: They require a short access delay, while the number of them is much fewer than that of type-2 devices.
- Type-2 devices: They do not have any constraint on access delay.

In general, type-1 and type-2 devices can be seen as delay-sensitive and delay-tolerant devices, respectively.

In order to support two different types of devices, two orthogonal RBs can be allocated. For each type of devices, a pool of preambles can be associated with one RB. This is the case to build two different access systems. Note that, with one RB, a pool of preambles can be dynamically divided into two sub-pools of preambles to support two types of devices with different probabilities of preamble collisions. In this case, in order to have a sufficiently large number of preambles, a wide system bandwidth might be required, which may result in a low spectral efficiency.

In fact, the access delay depends on not only preamble collisions but also preamble detection errors,[3] since an active device may retransmit another preamble if the preamble transmitted previously is not detected (due to either collision or detection error). Thus, for a short access delay, both the probabilities of preamble collisions and detection errors have to be low. This implies that it is also necessary to take into account the probability of preamble detection errors to support different priorities. To this end, based on the notion of power-domain NOMA, we can design layered preambles with one RB to support different priorities between type-1 and type-2 devices with a high spectral efficiency in terms of the probability of preamble detection errors. In this design, for different priorities, it is necessary to ensure type-1 devices have a better performance of preamble detection than type-2 devices, which leads to a short access delay.

6.4.2 Design of Layered Preambles

Suppose that one RB is allocated to support two different types of devices. Let N be the length of preamble sequences for a given RB. Since the system bandwidth is proportional to N, as long as N is fixed, the system bandwidth is fixed. If all the preambles are orthogonal, the total number of preambles, denoted by L, is equal to N. However, if non-orthogonal preambles are allowed, L can be larger than N. In particular, we consider Alltop sequences for non-orthogonal preambles as an example, while different sequences can also be used (e.g. Zadoff–Chu sequences (Chu, 1972)). With Alltop sequences, we have $L = N^2$ for a prime $N \geq 5$ (Foucart and Rauhut, 2013).

Let $\mathbf{x}_l$ denote the lth Alltop sequence of length N with $||\mathbf{x}_l|| = 1$ for all l. Denote by L_i for $i \in \{1,2\}$ the number of preamble sequences assigned to type-i devices. Let $\mathcal{L}_1 = \{\mathbf{x}_1, \ldots, \mathbf{x}_N\}$ be the set of preambles for type-1 devices with $L_1 = N$. In addition, let $\mathcal{L}_2 = \{\mathbf{x}_{N+1}, \ldots, \mathbf{x}_{N+L_2}\}$ denote the set of preambles for type-2 devices with $L_2 \leq N(N - 1)$. For convenience, the lth preambles of $\mathcal{L}_1$ and $\mathcal{L}_2$ are denoted by $\mathbf{c}_l$ and $\bar{\mathbf{c}}_l$ and referred to as type-1 and type-2 preambles, respectively. Thus, we assume that $\mathcal{L}_1$ is a set of orthogonal preambles. On the other hand, $\mathcal{L}_2$ is a set of non-orthogonal preambles. In addition, since type-1 devices have a higher priority than type-2 devices, type-1 preambles can have a high signal power than type-2 preambles as shown in Figure 6.7. That is, the receive power of type-1 preambles, denoted by P_1, has to be greater than that of type-2 preambles P_2, i.e. $P_1 > P_2$. The resulting random access scheme with two different preamble sets is referred to as

---

3 There can be detection errors due to channel fading and/or the noise at the BS.

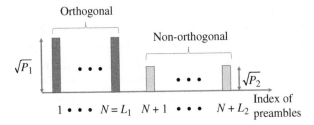

Figure 6.7 Two sets of preambles for type-1 and type-2 devices with different power levels for RALP.

Table 6.1 Absolute values of cross-correlation of layered preambles, $\mathcal{L}_1 = \{c_l\}$ and $\mathcal{L}_2 = \{\bar{c}_l\}$.

$\|\mathbf{x}_n^H \mathbf{x}_i\|$	$\mathbf{c}_1$	$\mathbf{c}_2$	$\cdots$	$\mathbf{c}_N$	$\bar{\mathbf{c}}_1$	$\bar{\mathbf{c}}_2$	$\cdots$	$\bar{\mathbf{c}}_{N(N-1)}$
$\mathbf{c}_1$	1	0	$\cdots$	0	$\frac{1}{\sqrt{N}}$	$\frac{1}{\sqrt{N}}$	$\cdots$	$\frac{1}{\sqrt{N}}$
$\mathbf{c}_2$	0	1	$\cdots$	0	$\frac{1}{\sqrt{N}}$	$\frac{1}{\sqrt{N}}$	$\cdots$	$\frac{1}{\sqrt{N}}$
$\vdots$	$\vdots$	$\vdots$	$\ddots$	$\vdots$	$\vdots$	$\vdots$	$\ddots$	$\vdots$
$\mathbf{c}_N$	0	0	$\cdots$	1	$\frac{1}{\sqrt{N}}$	$\frac{1}{\sqrt{N}}$	$\cdots$	$\frac{1}{\sqrt{N}}$
$\bar{\mathbf{c}}_1$	$\frac{1}{\sqrt{N}}$	$\frac{1}{\sqrt{N}}$	$\cdots$	$\frac{1}{\sqrt{N}}$	1	$\frac{1}{\sqrt{N}}$	$\cdots$	$\frac{1}{\sqrt{N}}$
$\bar{\mathbf{c}}_2$	$\frac{1}{\sqrt{N}}$	$\frac{1}{\sqrt{N}}$	$\cdots$	$\frac{1}{\sqrt{N}}$	$\frac{1}{\sqrt{N}}$	1	$\cdots$	$\frac{1}{\sqrt{N}}$
$\vdots$	$\vdots$	$\vdots$	$\ddots$	$\vdots$	$\vdots$	$\vdots$	$\ddots$	$\vdots$
$\bar{\mathbf{c}}_{N(N-1)}$	$\frac{1}{\sqrt{N}}$	$\frac{1}{\sqrt{N}}$	$\cdots$	$\frac{1}{\sqrt{N}}$	$\frac{1}{\sqrt{N}}$	$\frac{1}{\sqrt{N}}$	$\cdots$	1

random access with layered preambles (RALP), as $\mathcal{L}_1$ and $\mathcal{L}_2$ can be seen as layered preamble sets with different power allocations.

For Alltop sequences, the correlation between any two non-orthogonal preambles is $\frac{1}{\sqrt{N}}$. Thus, the correlation between any two different preambles in $\mathcal{L}_2$ is $\frac{1}{\sqrt{N}}$, while that in $\mathcal{L}_1$ is zero. In Table 6.1, the absolute values of the cross-correlations of layered preambles, $\mathcal{L}_1 = \{c_l\}$ and $\mathcal{L}_2 = \{\bar{c}_l\}$, are shown.

Since orthogonal sequences are used for $\mathcal{L}_1$, the detection of type-1 preambles can be carried out using the outputs of correlators. Each output of the correlator with a type-1 preamble, i.e. $\mathbf{c}_l$, does not have any interference from the other type-1 preambles due to the orthogonality. However, there is interference from type-2 preambles, as the correlation between one in $\mathcal{L}_1$ and another one in $\mathcal{L}_2$ is $\frac{1}{\sqrt{N}}$. It is noteworthy that although the number of type-2 preambles, $L_2 = |\mathcal{L}_2|$, can be as large as $N(N-1)$ and a large L_2 can reduce preamble collision for type-2 devices, a large L_2 may not be desirable in terms of the complexity and performance of the preamble detection at the BS.

6.4.3 Performance Analysis

To see the performance for type-1 devices, the following two metrics can be considered: (i) probability of preamble collision and (ii) outage probability of SINR (signal-to-interference-plus-noise ratio). To find those probabilities, we assume that the numbers of active type-1 and type-2 devices, follow independent Poisson distributions with parameters, λ_1 and λ_2, respectively.

For given K_1 active type-1 devices, the conditional probability of no preamble collision, which is equivalent to the probability that each bin has at most 1 ball when K_1 balls are independently thrown into L_1 bins, can be found as

$$\mathbb{P}_1(K_1) = \prod_{k=1}^{K_1-1} \left(1 - \frac{k}{L_1}\right) \approx e^{-\frac{K_1(K_1-1)}{2L_1}}, \tag{6.25}$$

where the approximation is actually a lower-bound (thus, $1 - e^{-\frac{K_1(K_1-1)}{2L}}$ is an upper-bound on the conditional probability of preamble collision). If K_1 is assumed to follow a Poisson distribution with mean λ_1 as mentioned earlier, it can be shown that

$$\begin{aligned}
\mathbb{P}_1 &= \mathbb{E}\left[\sum_{K_1=0}^{\infty} e^{-\frac{K_1(K_1-1)}{2L_1}}\right] \\
&\approx \mathbb{E}\left[\sum_{K_1=0}^{\infty}\left(1 - \frac{K_1(K_1-1)}{2L_1}\right)\right] \\
&= 1 - \frac{\lambda_1^2}{2L_1}.
\end{aligned} \tag{6.26}$$

Thus, $\frac{\lambda_1^2}{2L_1}$ becomes the probability of preamble collision. It can be shown that

$$1 - \mathbb{P}_1 \leq \delta \Rightarrow \frac{\lambda_1^2}{2L_1} \leq \delta, \tag{6.27}$$

where δ is a threshold probability of preamble collision. In general, it is expected that δ is sufficiently low so that type-1 devices can have a low access probability.

To keep a target probability of preamble collision δ, we need

$$\lambda_1 \leq \sqrt{2\delta L_1} = \sqrt{2\delta N}. \tag{6.28}$$

Thus, the BS needs to ensure that the average number of active type-1 devices is less than or equal to $\sqrt{2\delta N}$.

Recall that it has been assumed that active devices can perform power-control to achieve a target receive power, $P_i, i \in \{1,2\}$. To this end, the channel reciprocity in time division duplexing (TDD) can be exploited. Let $\mathcal{K}_i$ denote the index set of active type-i devices. Note that $K_i = |\mathcal{K}_i|$. Then, the received signal at the BS when active devices transmit preambles is given by

$$\mathbf{y} = \sqrt{P_1} \sum_{k \in \mathcal{K}_1} \mathbf{c}_{l(k)} v_k + \sqrt{P_2} \sum_{k \in \mathcal{K}_2} \bar{\mathbf{c}}_{\bar{l}(k)} \bar{v}_k + \mathbf{n}, \tag{6.29}$$

where v_k and $\bar{v}_k$ represent the channel coefficients from the kth active type-1 and type-2 devices, respectively, and $\mathbf{n} \sim \mathcal{CN}(0, N_0\mathbf{I})$ is the background noise. In (6.29), due to the power control, we assume that $\mathbb{E}[|v_k|^2] = \mathbb{E}[|\bar{v}_k|^2] = 1$ and $l(k)$ and $\bar{l}(k)$ stand for the preamble indices of the kth active type-1 and type-2 devices, respectively.

Taking advantage of the orthogonality of $\mathcal{L}_1$ (i.e. the preambles for type-1 devices) and low interference from transmitted type-2 preambles of power P_2, the BS can detect them using the following correlator's output:

$$\begin{aligned}
g_l &= \mathbf{c}_l^H \mathbf{y} \\
&= \sqrt{P_1} x_l + \sqrt{P_2} \sum_{k \in \mathcal{K}_2} \mathbf{c}_l^H \bar{\mathbf{c}}_{\bar{l}(k)} \bar{v}_k + n_l, \quad l = 1, \ldots, L_1,
\end{aligned} \tag{6.30}$$

where $n_l = \mathbf{c}_l^H \mathbf{n}$ and $x_l = \sum_{k \in \mathcal{K}_1} v_k \mathbb{1}(l(k) = l)$. Clearly, due to the orthogonality of $\mathcal{L}_1$, there is no interference from the other active type-1 devices of high receive power. If there is only one type-1 device that chooses $\mathbf{c}_l$, the SINR from (6.30) can be given by $\mathsf{SINR}_1 = \frac{P_1}{P_2 \frac{K_2}{N} + N_0}$. Then, it can be seen that the SINR of active type-1 device depends on the number of active type-2 devices. The outage probability of SINR can be defined as

$$\begin{aligned}
\mathbb{P}_{\text{out},1} &= \Pr(\mathsf{SINR}_1 \leq \Gamma_1) \\
&= \Pr\left(K_2 \geq \frac{N}{\gamma_2}\left(\frac{\gamma_1}{\Gamma_1} - 1\right)\right),
\end{aligned} \tag{6.31}$$

where $\gamma_i = \frac{P_i}{N_0}$. Thus, for a fixed outage of SINR, it can be seen that λ_2 can grow linearly with N.

In Figure 6.8, we show the values of (λ_1, λ_2) when $N = L_1$ varies to keep $\delta = \mathbb{P}_{\text{out},1} = 0.01$ when $\gamma_1 = 12$ dB, $\gamma_2 = 6$ dB, and $\Gamma_1 = 5$ dB. As N increases, the average numbers of active devices, λ_1 and λ_2, can increase for given δ and $\mathbb{P}_{\text{out},1}$.

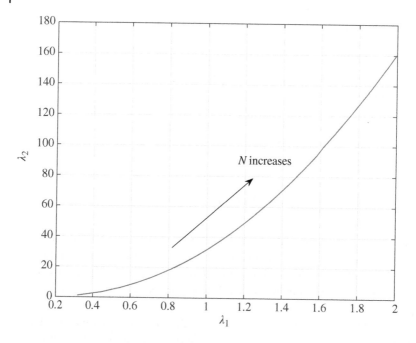

Figure 6.8 (λ_1, λ_2) for various values of N to keep $\delta = \mathbb{P}_{out,1} = 0.01$.

Although λ_1 and λ_2 are decided for type-1 devices to keep the probability of preamble collision and outage probability of SINR low as shown above, it is inevitable to have preamble detection errors, e.g. false alarm (FA) and miss detection (MD) errors. FA errors are the detection errors when the receiver detects preambles that are not transmitted, while MD errors are those when the receiver fails to detect transmitted preambles. MD errors in the type-1 preamble detection result in a high interference. On the other hand, FA errors result in the dimension reduction. Thus, MD errors have a worse performance degradation of type-2 preamble detection than FA errors.

After the type-1 preamble detection, SIC can be carried out for the type-2 preamble detection. For convenience, let $\mathcal{J}_i$ denote the index set of the preambles that are chosen by active type-i devices, i.e. $\mathcal{J}_i = \{l : K_{i,l} \geq 1\}$, where $K_{i,l}$ represents the number of active type-i devices that choose preamble l. Note that if there are no errors in type-1 preamble detection, $\mathcal{J}_1 = \{l(1), \ldots, l(K_1)\}$. For $l \in \mathcal{J}_1$, we assume that $\frac{1}{\sqrt{P_1}} g_l$ is an estimate of s_l. Thus, for SIC, the received signal from active type-1 devices can be reconstructed and removed as follows:

$$\bar{\mathbf{y}} = \mathbf{y} - \sum_{l \in \mathcal{J}_1} g_l \mathbf{c}_l$$

$$= \mathbf{P}_1 \left(\sqrt{P_2} \sum_{k \mathcal{K}_2} \bar{\mathbf{c}}_l \bar{s}_l + \mathbf{n} \right), \tag{6.32}$$

where $\mathbf{P}_1 = \mathbf{I} - \sum_{l \in \mathcal{J}_1} \mathbf{c}_l \mathbf{c}_l^H$ is an orthogonal projection matrix. This implies that SIC results in the signal suppression that suppresses all the signals in the subspace spanned by $\mathbf{c}_l, l \in \mathcal{J}_1$.

For the type-2 preamble detection, we can extend the case that the BS has a number of antennas, say M antennas, to improve detection performance via diversity combining or beamforming. As a result, $\mathbf{g}_l$ stands for the estimate of the channel vector of the active type-1 device that chooses the lth type-1 preamble, $\mathbf{c}_l$.

Once all the preambles transmitted from active type-1 devices are detected and removed using SIC, the type-2 preamble detection is reduced to as a sparse signal recovery problem. There are a number of approaches to sparse signal recovery (Eldar and Kutyniok, 2012). Among those, we can use an approach that is widely used for statistical inference, namely the coordinate ascent variational inference (CAVI) algorithm (Blei et al., 2017), which has been successfully used in Choi (2019b) to detect sparse signals in MTC.

We present simulation results when Alltop sequences of length $N = 13$ are used. In particular, we focus on results of the type-2 preamble detection using the CAVI algorithm with five iterations. To see the impact of error propagation due to MD and FA events in the type-1 preamble detection, we consider two different cases: (i) one preamble in $\mathcal{J}_1^c$ is incorrectly detected (i.e. an event of FA) and (ii) one preamble in $\mathcal{J}_1$ is not detected (i.e. an event of MD). Furthermore, we assume that

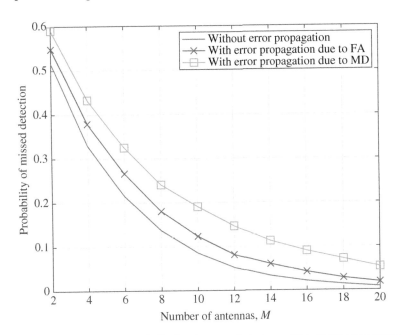

Figure 6.9 Probabilities of MD of active type-2 devices with/without error propagation (due to FA and MD in type-1 preamble detection) as functions of M when $K_2 = 5$, $N = L_1 = 13, L_2 = 5N, \gamma_1 = 12$ dB, and $\gamma_2 = 6$ dB.

the BS knows the number of active type-2 devices, K_2. Thus, we only consider the number of MD events, which is the same.

In Figure 6.9, the probabilities of MD of active type-2 devices with/without error propagation are shown as functions of the antennas at the BS, M, when $K_2 = 5$, $N = L_1 = 13$, $L_2 = 5N$, $\gamma_1 = 12$ dB, and $\gamma_2 = 6$ dB. It is shown that a large number of antennas, M, is desirable for a reasonable performance of type-2 preamble detection with/without error propagation.

It is expected that the probability of preamble collision decreases with L_2 when K_2 is fixed. However, if L_2 increases, the complexity of the CAVI algorithm increases and its performance is also degraded. To see the impact of L_2 on the performance of the type-2 preamble detection, we show the probabilities of MD of active type-2 devices with/without error propagation in Figure 6.10 as functions of the size of the preamble pool for type-2 devices, L_2, when $K_2 = 5$, $N = L_1 = 13$, $M = 10$, $\gamma_1 = 12$ dB, and $\gamma_2 = 6$ dB. As expected, the probability of MD increases with L_2. From this, we can see that there is a trade-off between the probability of preamble collision and the probability of MD, and L_2 should be chosen for a balanced performance in terms of both probabilities.

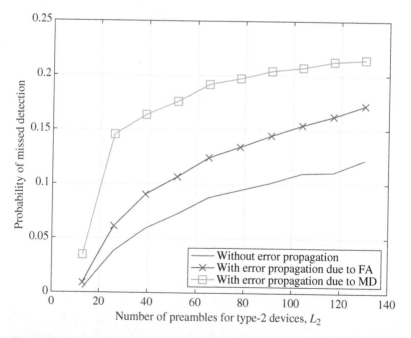

Figure 6.10 Probabilities of MD of active type-2 devices with/without error propagation (due to FA and MD in type-1 preamble detection) as functions of L_2 when $K_2 = 5$, $N = L_1 = 13$, $M = 10$, $\gamma_1 = 12$ dB, and $\gamma_2 = 6$ dB.

6.5 Further Reading

For two-step random access in 5G MTC standards, the reader is referred to Kim et al. (2021). In Choi et al. (2021), various extensions of two-step random access can also be found.

The notion of massive MIMO was presented for cellular systems in Marzetta (2010). Since then, massive MIMO has been extensively studied, and an overview can be found in Lu et al. (2014). Some examples of two-step random access with massive MIMO can be found in Björnson et al. (2017), Ding et al. (2020b), Choi (2021), and Choi and Ding (2021).

7

Game-Theoretic Perspective of NOMA-Based Random Access

Random access is a distributed access mechanism where each user is to compete for access on a shared channel. From this, users can be seen as players in a non-cooperative game, and game theory can play a crucial role in explaining users' behaviors in random access. In this chapter, we present basic notions of (non-cooperative) game theory and apply them to random access game. Furthermore, NOMA-based random access is analyzed from a game-theoretic perspective.

7.1 Background of Game Theory

In this section, we briefly review key elements of non-cooperative game theory that can allow to model users in random access contend for accessing shared channel resources as players in a non-cooperative game.

7.1.1 Normal-Form Games

In a game, there are multiple players and each player has a set of actions or strategies to choose. In addition, each player receives a payoff that depends the strategies chosen by the player as well as the other players. Hence, the normal-form or strategic-form representation of a game specifies:

(i) the set of players, denoted by $\mathcal{K} = \{1, \ldots, K\}$;
(ii) the sets of strategies or actions of players, denoted by S_k for player k;
(iii) the payoff or utility functions of players, denoted by u_k for player k.

Thus, a K-player normal-form game can be represented by

$$G = \{\mathcal{K}, S_1, \ldots, S_K; u_1, \ldots, u_K\}.$$

Massive Connectivity: Non-Orthogonal Multiple Access to High Performance Random Access,
First Edition. Jinho Choi.
© 2022 The Institute of Electrical and Electronics Engineers, Inc. Published 2022 by John Wiley & Sons, Inc.

A normal-form game with a finite number of strategies is called a finite normal-form game.

As mentioned earlier, since the payoff received by player k depends on all players' chosen strategies, it is a function of s_k as well as $\mathbf{s}_{-k}$, i.e. $u_k = u_k(s_k, \mathbf{s}_{-k})$, where $s_k \in S_k$ represents the strategy chosen by player k and $\mathbf{s}_{-k} = \{s_1, \ldots, s_{k-1}, s_{k+1}, \ldots, s_K\}$ stands for the set of the other players' chosen strategies.

Example 7.1 Consider a two-player game (i.e. $K = 2$). Each player has a coin and chooses either a head or a tail of this coin. The players then reveal their choices simultaneously. If both heads or both tails, then Player 1 keeps both coins, so wins one from Player 2. Otherwise (i.e. one head and one tail), Player 2 wins and takes both coins. In this game, we have $S_k = \{H, T\}$ for $k = 1, 2$, and

$$u_1(s_1, s_2) = \begin{cases} 1, & \text{if } s_1 = s_2, \\ -1, & \text{otherwise} \end{cases}$$

$$u_2(s_2, s_1) = \begin{cases} -1, & \text{if } s_1 = s_2, \\ 1, & \text{otherwise} \end{cases}$$

The resulting game is called matching pennies, which is a bimatrix game that has two matrices for the payoffs to two users as shown in Table 7.1 (the first element is for the 2×2 payoff matrix of Player 1, while the second for the 2×2 payoff matrix of Player 2).

The sum of the payoffs at each outcome is zero, and this class of games is called a zero-sum game, in which gains to one player can occur only at the expense of losses to another player.

7.1.2 Nash Equilibrium

Suppose that player k wants to choose s_k that maximizes payoff $u_k(s_k, \mathbf{s}_{-k})$. Thus, for given $\mathbf{s}_{-k}$, the best response of player k can be defined as

$$\text{BR}(\mathbf{s}_{-k}) = \underset{s_k \in S_k}{\arg\max}\, u_k(s_k, \mathbf{s}_{-k}), \tag{7.1}$$

Table 7.1 Bimatrix of matching pennies game (Players 1 and 2 are referred to as row and column players, respectively).

Player 1 \ Player 2	H	T
H	$(1, -1)$	$(-1, 1)$
T	$(-1, 1)$	$(1, -1)$

which shows that the payoff also depends on the other players' chosen strategies, i.e. the best response of player k is a function of $\mathbf{s}_{-k}$. As a result, it is not possible for player k to choose the strategy maximizing the payoff without knowing the others' chosen strategies. Thus, each player can choose the best response to the predicted strategies of the other players, which leads to the notion of Nash equilibrium (NE). In particular, the K-player normal-form game $G = \{S_1, \ldots, S_K; u_1, \ldots, u_K\}$, the strategies $\mathbf{s}^* = \{s_1^*, \ldots, s_K^*\}$ are NE, if

$$u_k(s_k^*, \mathbf{s}_{-k}^*) \geq u_k(s_k, \mathbf{s}_{-k}^*), \text{ for all } s_k \in S_k \text{ and } k \in \{1, \ldots, K\}. \tag{7.2}$$

Thus, no player has an incentive to deviate from a NE.

Example 7.2   Consider the game in Example 7.1. If Player 2 chooses $s_2 = H$, the best response of Player 1 is $s_1 = H$, because $u_1(H, H) = 1 > u_1(T, H) = -1$. This game does not have a NE because there is no strategy that is a best response to a best response.

7.1.3   Mixed Strategies

So far, it was assumed that each player can choose a strategy from S_k. It is possible to extend it with randomization. Let $\sigma_k(s_k)$ be the probability that s_k is chosen by player k. Thus, $\sigma_k(s_k) \geq 0$ and

$$1 = \sum_{s_k \in S_k} \sigma_k(s_k). \tag{7.3}$$

Then, the probability distribution $\sigma_k = \{\sigma_k(s_k), s_k \in S_k\}$ is called a mixed strategy, while s_k is referred to as a pure strategy. Let $L_k = |S_k|$, i.e. L_k represents the number of the strategies that player k has. In addition, let Σ_L denote the L-dimensional simplex that is defined by

$$\Sigma_L = \left\{ [p_0 \ p_1 \ \cdots \ p_L]^T \ : \ 0 \leq p_l \leq 1, \ \sum_{l=0}^{L} p_l = 1 \right\},$$

which is a convex set. Then, $\sigma_k \in \Sigma_{L_k-1}$. The utility function becomes a function of $\{\sigma_1, \ldots, \sigma_K\}$ as follows:

$$u_k(\sigma_1, \ldots, \sigma_K) = \sum_{s_1 \in S_1} \cdots \sum_{s_K \in S_K} u_k(s_1, \ldots, s_K) \sigma_1(s_1) \cdots \sigma_K(s_K). \tag{7.4}$$

With mixed strategies, the best response can be generalized as follows:

$$\mathrm{BR}_k(\sigma_{-k}) = \operatorname*{argmax}_{\sigma_k \in \Sigma_{L_k-1}} u_k(\sigma_k, \sigma_{-k}) \in \Sigma_{L_k-1}. \tag{7.5}$$

For $\sigma = (\sigma_1, \ldots, \sigma_K) \in \Sigma = \prod_{k=1}^K \Sigma_{L_k-1}$, let

$$\mathrm{BR}(\sigma) = (\mathrm{BR}_1(\sigma_{-1}), \ldots, \mathrm{BR}_K(\sigma_{-K})) \in \Sigma. \tag{7.6}$$

If a $\sigma \in \Sigma$ satisfies

$$\sigma \in BR(\sigma),$$

then it is a fixed point of $BR(\cdot)$. A set of the mixed strategies, denoted by $\{\sigma_1^*, \ldots, \sigma_K^*\}$, is a mixed strategy NE if

$$u_k(\sigma_k^*, \sigma_{-k}^*) \geq u_k(\sigma_k, \sigma_{-k}^*), \text{ for all } \sigma_k \text{ and } k \in \{1, \ldots, K\}. \tag{7.7}$$

It is known that a mixed strategy NE exists for any finite normal-form game (Fudenberg and Tirole, 1991; Osborne and Rubinstein, 1994). Clearly, a mixed strategy NE is a fixed point of $BR(\cdot)$, and the properties of the best response are used to prove that there exists a mixed strategy NE for any finite normal-form game.

Example 7.3 Consider the game in Example 7.1. As shown earlier, this game does not have a pure strategy NE, but a mixed strategy NE. Let $p = \sigma_1(H)$ and $q = \sigma_2(H)$. Then, it can be shown that

$$u_1(p, q) = pq + (1 - p)(1 - q) - p(1 - q) - (1 - p)q = (1 - 2p)(1 - 2q),$$

while $u_2(q, p) = -u_1(p, q)$. The mixed strategy NE is $(p, q) = \frac{1}{2}, k = 1, 2$.

7.2 Random Access Game

In this section, we can consider a random access game with $K = 2$ players (Felegy-hazi and Hubaux, 2006). Unlike conventional random access, in the random access game, users as players can choose to transmit or not even if they have packets to send. The main concern of users is to maximize their own payoffs, where the payoff is decided by the reward of successful transmission and the cost of transmission.

7.2.1 Normal-Form and NE

In the random access game, each player has the set of actions or strategies, $S_k = \{T, Q\}$, where T and Q stand for "transmit" and "stay quite," respectively. We assume that player k can receive a unit reward if the transmission is successful without collision. Then, the payoff can be decided as follows:

$$u_k(T, T) = -c, \quad u_k(T, Q) = 1 - c, \quad u_k(Q, T) = 0, \quad \text{and} \quad u_k(Q, Q) = 0,$$

where $c > 0$ is the cost of transmission. For example, if Players 1 and 2 choose to transmit, a collision occurs and the player has no reward, only a cost. As a result, $u_k(T, T) = -c$ for $k \in \{1, 2\}$. The resulting game is symmetric and the bimatrix in Table 7.2 can be found.

Table 7.2 Bimatrix of two-person random access game.

Player 1 \ Player 2	Q	T
Q	$(0, 0)$	$(0, 1 - c)$
T	$(1 - c, 0)$	$(-c, -c)$

For the random access game with two players, $(s_1, s_2) = (Q, T)$ and (T, Q) are NE for $0 < c < 1$. To see this, consider $(s_1, s_2) = (Q, T)$. It can be shown that

$$u_1(Q, T) = 0 \geq u_1(T, T) = -c,$$
$$u_2(T, Q) = 1 - c \geq u_2(Q, Q) = 0,$$

which shows that $(s_1, s_2) = (Q, T)$ is an NE.

7.2.2 Mixed Strategies

In the random access game, let p_k be the probability that strategy T is chosen by player k. Then, it can be shown that

$$u_k(T, p_{-k}) = 1 - c - p_{-k},$$
$$u_k(Q, p_{-k}) = 0. \tag{7.8}$$

Here, for $k = 1$ and 2, $p_{-k} = p_2$ and $p_{-k} = p_1$, respectively. Thus, in terms of p_1 and p_2, the payoff function is given by

$$u_k(p_k, p_{-k}) = p_k(1 - c - p_{-k}). \tag{7.9}$$

As a result, the best response can be found as follows:

$$BR(p_{-k}) = \begin{cases} 1, & \text{if } p_{-k} < 1 - c, \\ [0, 1], & \text{if } p_{-k} = 1 - c, \\ 0, & \text{if } p_{-k} > 1 - c. \end{cases} \tag{7.10}$$

Note that if $p_{-k} = 1 - c$, $u_k(p_k, p_{-k}) = 0$, which is independent of p_k. That is, the best response becomes any value of p_k between 0 and 1. Thus, the fixed point of $BR(p_1, p_2)$ is $p_1 = p_2 = 1 - c$, because $p_k = 1 - c \in BR(p_{-k} = 1 - c)$. This is also the mixed strategy NE of the random access game.

7.3 NOMA-ALOHA Game

As shown by the random access game in Section 7.2, the notion of game theory can be well applied to random access where users compete for a shared channel. In this section, we extend the random access game based on the notion of power-domain NOMA that can improve the throughput.

7.3.1 Single-Channel NOMA-ALOHA Game

Suppose that there are two users (throughout this section, we assume that users and players are interchangeable) and a receiver or base station (BS) for multiple access (or uplink transmissions) with a single channel. We assume that NOMA is employed with two different power levels. Thus, the number of transmission strategies for each user is three as follows:

$$S_k = S = \{H, L, 0\}, \quad l = 1, 2,$$

where the subscript k is the index for users, H and L represent the high and low transmit powers, respectively, and 0 represents no transmission. Denote by s_k the strategy of user k. In addition, $\mathbf{s}_{-k}$ represents the set of the strategies of the users except user k, i.e. $\mathbf{s}_{-k} = \{s_1, \ldots, s_{k-1}, s_{k+1}, \ldots, s_K\}$, if there are K users. For two-person games, i.e. $K = 2$, $\mathbf{s}_{-1} = s_2$ and $\mathbf{s}_{-2} = s_1$.

The payoff function of user k is given by

$$u_k(s_k, \mathbf{s}_{-k}) = u(s_k, \mathbf{s}_{-k})$$
$$= R(s_k, \mathbf{s}_{-k}) - C(s_k), \quad k = 1, 2, \tag{7.11}$$

where $R(s, s')$ is the reward function of successful transmission and $C(s)$ is the cost function of transmission strategy. In particular, we consider the following reward function:

$$R(s, s') = \begin{cases} W, & \text{if } s \neq s' \text{ and } s \in \{H, L\}, \\ 0, & \text{otherwise,} \end{cases} \tag{7.12}$$

where $W > 0$ is the reward of successful transmission for a user. Due to NOMA, if the user of interest chooses $s = H$, while the other user chooses $s' = L$ or 0, the user of interest can successfully transmit his signal. Thus, the main difference of NOMA-ALOHA from conventional ALOHA is that the BS is able to recover the signals from two users simultaneously as long as one user employs strategy H and the other user adopts strategy L as in (7.12). Note that in (7.12), as in conventional ALOHA, if two users choose (H, H) or (L, L), we assume collision and the BS is not able to receive any signal.

For the cost function, we can consider the following assignment as an example:

$$C(H) = 2, \quad C(L) = 1, \quad \text{and} \quad C(0) = 0,$$

because strategy H requires a higher transmit power than strategy L.

It is noteworthy that the payoff function in (7.11) can be seen as an energy-efficiency metric, which is widely used in wireless systems, e.g. Saraydar et al. (2002) for power control game. To see that the payoff function in (7.11) is an energy-efficiency metric, we can consider the logarithm of the ratio of the spectral efficiency or throughput to the transmit power as follows:

$$\ln \frac{\text{Throughput}}{\text{Transmit Power}} = \ln(\text{Throughput}) - \ln(\text{Transmit Power}),$$

where $\ln(\text{Throughput})$ becomes the reward function and $\ln(\text{Transmit Power})$ becomes the cost function in (7.11).

The resulting game is symmetric, that is, both the players have the same set of strategies, and their payoff functions satisfy $u_1(s_1, s_2) = u_2(s_2, s_1)$ for each $s_1, s_2 \in S$. Furthermore, its bimatrix can be found as in Table 7.3.

It is noteworthy that the two-person NOMA-ALOHA game can be seen as a generalization of the random access game with the notion of NOMA. The random access game has two strategies for each user, namely Transmit (T) and Quite (Q). Strategy T is further divided into H and L in the two-person NOMA-ALOHA game, while strategy Q becomes 0.

The two-person NOMA-ALOHA game has NEs that depend on the reward, W. If $W \leq 2$, there exist pure strategy NEs, denoted by $\{s_k^*\}$, which are characterized by

$$u_k(s_k^*, s_{-k}^*) \geq u_k(s_k, s_{-k}^*), \quad \text{for all } s_k \in S_k, \ k \in \mathcal{K}.$$

For example, for $0 \leq W < 1$, from Table 7.3, we can see that $(s_1, s_2) = (0, 0)$ is the pure strategy NE. That is, if the reward of successful transmission, W, is sufficiently small (compared to the cost of transmissions), the users do not want to transmit signals and non-transmission strategy (i.e. $s_k = 0$) becomes NE. For $1 \leq W \leq 2$, the pure strategy NEs are $(s_1, s_2) = (0, L)$ and $(s_1, s_2) = (L, 0)$. If $W > 2$, there is no pure strategy NE.

In general, we are interested in mixed strategy NEs as randomized strategy can be well employed for random access. In order to find mixed strategy NEs,[1] the principle of indifference (Maschler et al., 2013) can be used. Since the two-person

Table 7.3 Bimatrix of two-person NOMA-ALOHA game.

Player 1 \\ Player 2	H	L	0
H	$(-2, -2)$	$(W-2, W-1)$	$(W-2, 0)$
L	$(W-1, W-2)$	$(-1, -1)$	$(W-1, 0)$
0	$(0, W-2)$	$(0, W-1)$	$(0, 0)$

1 We only consider an approach to find mixed strategy NEs, while there are a number of different approaches (see (Fudenberg and Tirole, 1991; Osborne and Rubinstein, 1994)).

NOMA-ALOHA game is symmetric, it suffices to find one user's mixed strategy NE. To this end, let a and b denote the probabilities to choose H and L, respectively. Thus, without the subscript k for the index of player, a mixed strategy is represented by $\sigma = (a, b, 1 - a - b)$, where $a + b \leq 1$.

For convenience, let **B** denote the payoff matrix for the row user in Table 7.3. Let $[\mathbf{B}]_{n,m} = B_{n,m}$. According to the principle of indifference, the row user has the same expected payoff for any pure strategy when the column user employs the mixed strategy NE. Thus, it follows

$$
\begin{aligned}
U &= a^* B_{1,1} + b^* B_{1,2} + (1 - a^* - b^*) B_{1,3} \\
&= a^* B_{2,1} + b^* B_{2,2} + (1 - a^* - b^*) B_{2,3} \\
&= a^* B_{3,1} + b^* B_{3,2} + (1 - a^* - b^*) B_{3,3},
\end{aligned}
\tag{7.13}
$$

where U is the expected payoff of the row user and $(p^*, q^*, 1 - p^* - q^*)$ is the mixed strategy NE. From (7.13), we can have two equations for two unknown variables, a^* and b^*. In addition, since $a^* + b^* \leq 1$, we can find a^* and b^*.

Noting that $B_{3,i} = 0$ for $i = 1, 2, 3$ (as the row user does not transmit) from Table 7.3, we can see that $U = 0$ if $1 - (a^* + b^*) > 0$ (i.e. the probability of non-transmission or strategy 0 is greater than 0). In this case, we can have closed-form expressions for a^* and b^* from (7.13) as follows:

$$
a^* = \frac{W - 2}{W} \quad \text{and} \quad b^* = \frac{W - 1}{W}.
\tag{7.14}
$$

The above solution is valid when $2 \leq W < 3$ since $a^*, b^* \geq 0$ and $1 - (a^* + b^*) > 0$ are required. If $W = 3$, we can see that $a^* + b^* = 1$, which means that the probability of non-transmission is 0. That is, strategy 0 is not used if the reward of successful transmission, W, is sufficiently large. Thus, for $W \geq 3$, (7.13) is reduced to

$$
\begin{aligned}
U &= a^* B_{1,1} + b^* B_{1,2} + (1 - a^* - b^*) B_{1,3} \\
&= a^* B_{2,1} + b^* B_{2,2} + (1 - a^* - b^*) B_{2,3}.
\end{aligned}
\tag{7.15}
$$

Then, after some manipulations, we have

$$
a^* = \frac{W - 1}{2W} \quad \text{and} \quad b^* = \frac{W + 1}{2W}, \qquad W \geq 3.
\tag{7.16}
$$

We now consider the case that $W < 2$. If $W < 2$, the reward of successful transmission is so small that high-power transmission is not desirable. Thus, $a^* = 0$ (which is the case that $W = 2$ as shown in (7.14)). Thus, (7.13) is reduced to

$$
U = b^* B_{2,2} + (1 - b^*) B_{2,3} = 0,
\tag{7.17}
$$

which leads to

$$
b^* = \begin{cases} \frac{W-1}{W}, & 1 \leq W \leq 2, \\ 0, & 0 \leq W < 1. \end{cases}
\tag{7.18}
$$

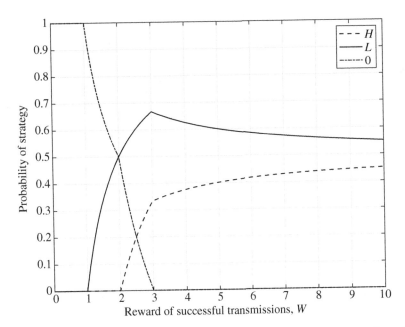

Figure 7.1 The mixed strategy NE, $\sigma^* = (a^*, b^*, 1 - a^* - b^*)$, for different values of the reward of successful transmission, W.

In Figure 7.1, we show the mixed strategy NE, $\sigma^* = (a^*, b^*, 1 - a^* - b^*)$, for different values of the reward of successful transmission, W. As shown in Figure 7.1, we can see that the probability of strategy L is higher than the probability of strategy H as strategy H has a higher cost than strategy L. In addition, as W increases, the probability of strategy 0 decreases. That is, as the reward of successful transmission increases, the users tend to transmit signals. Note that as $W \to \infty$, $a^* = b^* \to \frac{1}{2}$ from (7.16), i.e. the users always transmit.

When users always transmit in conventional ALOHA, there are collisions with probability (w.p.) 1 and the throughput (i.e. the average number of successfully transmitted packets) becomes 0. However, in NOMA-ALOHA, the throughput does not approach 0 although collisions happen thanks to NOMA. In the two-person NOMA-ALOHA, as $W \to \infty$, the asymptotic throughput approaches 1, because the BS is able to recover the two users' signals simultaneously as long as $(s_1, s_2) = (H, L)$ or (L, H), i.e.

$$\text{Throughput} = 2 \times \Pr\left((s_1, s_2) = (H, L) \text{ or } (L, H)\right)$$
$$= 2 \times \frac{1}{2} = 1, \quad W \to \infty.$$

Suppose that a mixed strategy is used and the two users have the same mixed strategy, $\sigma = (a, b, 1 - a - b)$, and use it independently (as no cooperation is

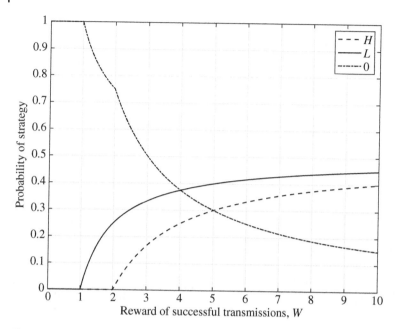

Figure 7.2 The mixed strategy that maximizes the average payoff for different values of the reward of successful transmission, W.

assumed). From (7.11), for a given mixed strategy, the average payoff is given by

$$\bar{u}(a,b) = \mathbb{E}[u(s_1,s_2)]$$
$$= \mathbb{E}[R(s_1,s_2)] - \mathbb{E}[C(s_1)]$$
$$= W\left(a(1-a) + b(1-b)\right) - 2a - b. \tag{7.19}$$

Since $\bar{u}(a,b)$ is concave in a and b, the maximization of the average payoff can be carried out. In Figure 7.2, we show the optimal mixed strategy, denoted by $\hat{\sigma} = (\hat{a}, \hat{b}, 1 - \hat{a} - \hat{b})$ that maximizes the average payoff for different values of the reward of successful transmission, W. We can see that the optimal mixed strategy that maximizes the average payoff is similar to the mixed strategy NEs in Figure 7.1, although both the mixed strategies are not the same for $W \geq 1$. For example, for $0 \leq W \leq 1$, the probability of strategy 0 is 1 in both the mixed strategies. In addition, we see that the probability of strategy L is higher than or equal to that of strategy H, while the two probabilities approach $\frac{1}{2}$ as $W \to \infty$ in both the mixed strategies.

Figure 7.3 shows the average payoff functions of the two mixed strategies, $\hat{\sigma}$ and σ^*, for different values of W. Clearly, $\hat{\sigma}$ provides the highest average payoff and higher than that of σ^*. However, since $\hat{\sigma}$ is not a mixed strategy NE, any user who

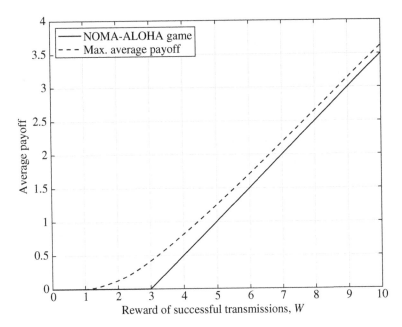

Figure 7.3 The maximum average payoff and the average payoff of the mixed strategy NE of the two-person NOMA-ALOHA for different values of W.

uses a slightly different mixed strategy from $\hat{\sigma}$ can have a higher average payoff at the cost of the degraded average payoff of the other user who uses $\hat{\sigma}$.

The ratio of the payoff with $\hat{\sigma}$ to that with σ^* can be seen as a price of anarchy (PoA) (Koutsoupias and Papadimitriou, 1999; Nisan et al., 2007). If both the users trust each other, they can employ $\hat{\sigma}$. On the other hand, if there is no trust, each user may need to employ σ^*, which is NE, and has a worse average payoff than that can be obtained with $\hat{\sigma}$, i.e. the PoA is less than 1. However, as $W \to \infty$, from (7.19), we can see that $\hat{a}$ and $\hat{b}$ become $\frac{1}{2}$, which is the same as the asymptotic mixed strategy NE, (a^*, b^*), with $W \to \infty$, and the PoA approaches 1.

Under symmetric conditions, the two-person NOMA-ALOHA game can be easily generalized with more than two players or users. Suppose that there are $K \geq 2$ users. Since each user has the same payoff function, the NOMA-ALOHA game is symmetric and the principle of indifference (Maschler et al., 2013) can be employed in order to find the mixed strategy NE.

For convenience, let the user of interest is user k. Denote by $U(H)$, $U(L)$, and $U(0)$ the payoff values of user 1 if user 1 chooses $s_k = H, L$, and 0, respectively, when the other users have the same mixed strategy, $(a, b, 1 - a - b)$. Let $q_k(s)$ be the probability that the other users do not employ strategy s. Since each user chooses a strategy independently, we have

$$
q_k(H) = \prod_{i \neq k} (1 - a) = (1 - a)^{K-1},
$$

$$
q_k(L) = \prod_{i \neq k} (1 - b) = (1 - b)^{K-1}. \tag{7.20}
$$

From this, we show that

$$
U(H) = -C(H)\left(1 - q_k(H)\right) + (W - C(H))q_k(H),
$$

$$
U(L) = (W - C(L))q_k(L) - C(L)\left(1 - q_k(L)\right), \tag{7.21}
$$

while the payoff of user k becomes $U(0) = 0$ when $s_k = 0$.

As shown earlier, we can see that if $W < 1$, both the probabilities of strategies H and L are to be 0 for NE. Thus, $(a^*, b^*, 1 - a^* - b^*) = (0, 0, 1)$. For $1 \leq W < 2$, strategy H cannot be applied for the mixed strategy NE (i.e. $a^* = 0$). In this case, since we need to have

$$
U(L) = U(0),
$$

the resulting mixed strategy NE becomes

$$
(a^*, b^*, 1 - a^* - b^*) = \left(0, 1 - \left(\frac{1}{W}\right)^{\frac{1}{K-1}}, \left(\frac{1}{W}\right)^{\frac{1}{K-1}}\right).
$$

To find the mixed strategy NE for $W \geq 2$, let

$$
W^* = \left(1 + 2^{\frac{1}{K-1}}\right)^{K-1}. \tag{7.22}
$$

Suppose that the probability that a user employs strategy 0 is not zero, while $a, b > 0$. Then, by the principle of indifference, we need to have $U(H) = U(L) = U(0) = 0$. From (7.21), it can be shown that $Wq_k(H) - 2 = Wq_k(L) - 1 = 0$. Thus, for $2 \leq W < W^*$, we have

$$
(a^*, b^*) = \left(1 - \left(1 - \frac{2}{W}\right)^{\frac{1}{K-1}}, 1 - \left(\frac{1}{W}\right)^{\frac{1}{K-1}}\right). \tag{7.23}
$$

However, if W is sufficiently large (i.e. for a sufficiently large reward of successful transmission), the probability that a user employs strategy 0 becomes zero or $a^* + b^* = 1$. The corresponding W is the solution of $1 = a^* + b^*$ or

$$
1 = \left(\frac{2}{W}\right)^{\frac{1}{K-1}} + \left(\frac{1}{W}\right)^{\frac{1}{K-1}}, \tag{7.24}
$$

which is W^* in (7.22). In this case (i.e. $W \geq W^*$), we only need to have $U(H) = U(L)$ with $a^* + b^* = 1$. Thus, we have

$$
Wq_k(H) - 2 = Wq_k(L) - 1 \quad \text{and} \quad a^* + b^* = 1,
$$

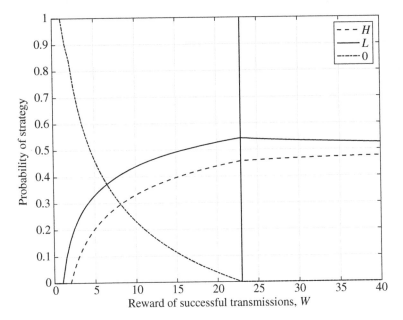

Figure 7.4 The mixed strategy NE, $\sigma^* = (a^*, b^*, 1 - a^* - b^*)$, for different values of the reward of successful transmission, W, when $K = 5$.

which can also be expressed in terms of b only as follows:

$$Wb^{K-1} = W(1 - b)^{K-1} + 1, \tag{7.25}$$

while $a^* = 1 - b^*$ for $W \geq W^*$. Clearly, the solution of (7.25) is b^*, and a^* becomes $1 - b^*$.

Figure 7.4 shows the mixed strategy NE, $\sigma^* = (a^*, b^*, 1 - a^* - b^*)$, for different values of the reward of successful transmission, W, when $K = 5$ with $W^* = 22.969$. Note that $W^* = 3$ when $K = 2$ according to (7.22), and W^* can be seen as the threshold value of the reward of successful transmission to set the probability of $s_k = 0$ to 0. Clearly, from (7.22), W^* increases with K. That is, a higher reward of successful transmission is required to force users to keep transmitting as K increases.

Figure 7.5 shows the mixed strategy NE, $\sigma^* = (a^*, b^*, 1 - a^* - b^*)$, for different numbers of users when $W = 10$. For a fixed reward of successful transmissions, as the number of users increases, the probability of transmissions (either H or L) decreases, while the probability of non-transmissions increases. This behavior results from the increase of the probability of collision as K increases.

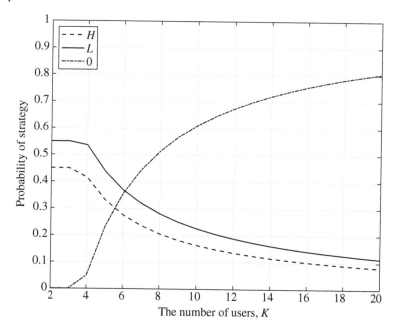

Figure 7.5 The mixed strategy NE, $\sigma^* = (a^*, b^*, 1 - a^* - b^*)$, for different numbers of users when $W = 10$.

7.3.2 Multichannel NOMA-ALOHA Game

NOMA can be applied to multichannel ALOHA with more shared channels, say N orthogonal channels, where each user can choose a channel uniformly at random and employ a strategy to transmit packets through the selected channel. Furthermore, as each user experiences independent fading in wireless channels, it is necessary to take into account different channel conditions for each user. In this case, due to different channel conditions, unlike the NOMA-ALOHA game in Section 7.3.1[GA1], the resulting NOMA-ALOHA game is no longer symmetric, which makes finding mixed strategy NE more involved (and a closed-form expression may not be available).

Suppose that there are N orthogonal channels and each user can choose one of them to transmit a packet with strategy H or L. A user may not transmit any packet if the cost of transmission is high, compared to the reward of successful transmission. Throughout this section, denote by $h_{k,n}$ the channel coefficient of channel n from user k to the BS. We assume that user k knows her channel state information (CSI), $\{h_{k,1}, \ldots, h_{k,N}\}$, but not the others' CSI.

Due to different channel conditions, each user has a different cost function that depends on CSI. In order to transmit a packet with strategy H or L, each user needs to decide the transmit power to overcome fading. If user k wants to transmit a packet with strategy H or L through channel n, the transmit power[2] becomes

$$P_{k,n} = \frac{P_H}{|h_{k,n}|^2} \quad \text{or} \quad \frac{P_L}{|h_{k,n}|^2},$$

respectively, where P_H and P_L represent the desired received power levels with strategies H and L, respectively. Then, the signal-to-noise ratio (SNR) at the BS becomes $\frac{P_H}{N_0}$ or $\frac{P_L}{N_0}$ with strategy H or L, respectively, as the received signal of channel n under the assumption that there is no other user transmitting through channel n is given by

$$r_n = h_{k,n} \sqrt{P_{k,n}} d_k + \eta_n,$$

where d_k is the transmitted signal from user k with $\mathbb{E}[|d_k|^2] = 1$ and $\eta_n \sim \mathcal{CN}(0, N_0)$ is the background noise of channel n. If there is another user, say user k', transmitting with strategy L through channel n when user k transmits with strategy H through channel n, the received signal becomes

$$r_n = h_{k,n} \sqrt{P_{k,n}} d_k + h_{k',n} \sqrt{P_{k',n}} d_{k'} + \eta_n.$$

From this, when the BS decodes the signal from user k, the signal-to-interference-plus-noise ratio (SINR) of user k of channel n can be given by

$$\gamma_{k,n} = \frac{P_H}{P_L + N_0}.$$

The power levels, P_H and P_L, are decided to guarantee successful decoding for a given code rate or transmission rate so that the SINR, $\frac{P_H}{P_L+N_0}$, can allow successful decoding for user k's signal. Once user k's signal is decoded, it is removed by successive interference cancellation (SIC). Then, user k''s signal can be decoded with SNR, $\frac{P_L}{N_0}$. In this case, all the signals from users k and k' can be successfully decoded. In particular, we assume that P_H and P_L are decided as follows:

$$\log_2 \left(1 + \frac{P_H}{P_L + N_0} \right) = \log_2 \left(1 + \frac{P_L}{N_0} \right) = r_{\text{tx}}, \tag{7.26}$$

where r_{tx} represents the achievable transmission rate.

---

2 Throughout this section, we consider that all the users in the game satisfy $\frac{P_H}{|h_{k,n}|^2} \leq P_{\max}$, where $P_{\max}$ is the maximum transmit power of a user. Thus, a user with a low channel coefficient that does not satisfy $\frac{P_H}{|h_{k,n}|^2} \leq P_{\max}$ cannot be active (in general, such a user is backlogged and waits until the channel gain is sufficiently high).

Denote by $a_{k,n}$ or $b_{k,n}$ the probability that user k transmits a packet with strategy H or L through channel n, respectively. Since only one channel is chosen at a time, we have

$$a_k = \sum_{n=1}^{N} a_{k,n} \leq 1, \ b_k = \sum_{n=1}^{N} b_{k,n} \leq 1,$$
$$\text{and } a_k + b_k \leq 1.$$

For convenience, let $\sigma_{k,n} = (a_{k,n}, b_{k,n}, 1 - a_{k,n} - b_{k,n})$ and $\sigma_k = [\sigma_{k,1} \ \cdots \ \sigma_{k,N}]^{\mathrm{T}}$. In addition, let Σ_k denote the set of all possible mixed strategy profiles for user k, σ_k. Since all the users have the same set, we simply write $\Sigma = \Sigma_1 = \cdots \ \Sigma_K$. For user k, the probabilities that strategies H and L through channel n are not used by the other users are

$$q_{k,n}(H) = \prod_{i \neq k} (1 - a_{i,n}) \quad \text{and} \quad q_{k,n}(L) = \prod_{i \neq k} (1 - b_{i,n}), \tag{7.27}$$

respectively. Then, for given mixed strategy, $\{\sigma_k\}$, the average reward becomes

$$\begin{aligned}
\bar{R}(\sigma_k, \sigma_{-k}) &= W \sum_{n=1}^{N} \left(a_{k,n} q_{k,n}(H) + b_{k,n} q_{k,n}(L) \right) \\
&= W \mathbf{q}_k^{\mathrm{T}} \mathbf{p}_k,
\end{aligned} \tag{7.28}$$

where

$$\begin{aligned}
\mathbf{p}_k &= [a_{k,1} \ \cdots \ a_{k,N} \ b_{k,1} \ \cdots \ b_{k,N}]^{\mathrm{T}}, \\
\mathbf{q}_k &= [q_{k,1}(H) \ \cdots \ q_{k,N}(H) \ q_{k,1}(L) \ \cdots \ q_{k,N}(L)]^{\mathrm{T}}.
\end{aligned} \tag{7.29}$$

The cost function depends on the CSI. Let user k's cost of transmission through channel n be

$$C_{k,n}(s_{k,n}) = \begin{cases} c_{k,n}(H), & \text{if } s_{k,n} = H, \\ c_{k,n}(L), & \text{if } s_{k,n} = L, \\ c_{k,n}(0), & \text{if } s_{k,n} = 0, \end{cases} \tag{7.30}$$

where $c_{k,n}(H) = \ln \frac{P_H}{|h_{k,n}|^2}$, $c_{k,n}(L) = \ln \frac{P_L}{|h_{k,n}|^2}$, and $c_{k,n}(0) = \ln \epsilon$. Here, in order to avoid a cost of $-\infty$ when no transmission is chosen, we consider a nominal positive power level ϵ for $c_{k,n}(0)$. In this section, we assume that $\epsilon = N_0$.

The average cost for given σ_k can be found as

$$\begin{aligned}
\bar{C}_k(\sigma_k) &= \sum_{n=1}^{N} c_{k,n}(H) a_{k,n} + c_{k,n}(L) b_{k,n} + c_{k,n}(0)(1 - a_{k,n} - b_{k,n}) \\
&= \mathbf{d}_k^{\mathrm{T}} \mathbf{p}_k + C_k(0),
\end{aligned} \tag{7.31}$$

where $C_k(0) = \sum_{n=1}^{N} c_{k,n}(0)$ and

$$
\mathbf{d}_k = \begin{bmatrix} c_{k,1}(H) - c_{k,1}(0) \\ \vdots \\ c_{k,N}(H) - c_{k,N}(0) \\ c_{k,1}(L) - c_{k,1}(0) \\ \vdots \\ c_{k,N}(L) - c_{k,N}(0) \end{bmatrix} = \begin{bmatrix} \ln \frac{P_H}{|h_{k,n}|^2 N_0} \\ \vdots \\ \ln \frac{P_H}{|h_{k,n}|^2 N_0} \\ \ln \frac{P_L}{|h_{k,n}|^2 N_0} \\ \vdots \\ \ln \frac{P_L}{|h_{k,n}|^2 N_0} \end{bmatrix}.
\tag{7.32}
$$

Finally, the payoff function is given by

$$
\begin{aligned}
\bar{u}_k(\sigma_k, \sigma_{-k}) &= \overline{R}_k(\sigma_k, \sigma_{-k}) - \overline{C}_k(\sigma_k) \\
&= W \mathbf{q}_k^{\mathrm{T}} \mathbf{p}_k - \mathbf{d}_k^{\mathrm{T}} \mathbf{p}_k - C_k(0).
\end{aligned}
\tag{7.33}
$$

Note that since $C_k(0)$ is independent of σ_k, we can remove $C_k(0)$ in (7.33) when finding the mixed strategy NE, $\{\sigma_k^*\}$, which is characterized by

$$
\bar{u}_k(\sigma_k^*, \sigma_{-k}^*) \geq \bar{u}_k(\sigma_k, \sigma_{-k}^*), \quad \text{for all } \sigma_k \in \Sigma.
\tag{7.34}
$$

Unfortunately, the resulting payoff function in (7.33) is not concave in $\mathbf{p}_k$. Since there can be distributed algorithms that can converge to the mixed strategy NE if the payoff function is concave, it is necessary to modify the payoff function so that it becomes concave. The interested reader is referred to Choi (2018a), in which a concave payoff function is obtained, and conditions for a unique NE are derived with a distributed algorithm that can converge to the NE.

7.4 Fictitious Play

In Sections 7.1–7.3, we discussed static games where all players choose actions simultaneously according to their mixed strategies. In addition, to find a mixed strategy NE of NOMA-ALOHA game, we have used the principle of indifference. While this is useful for analysis, it does not give any idea how an actual distributed player might find a mixed strategy NE.

In this section, we consider a different setting in which each player can observe the actions of the other players in the past and choose an action to maximize the payoff based on the prediction of the distribution of the others' actions. The resulting approach is referred to as fictitious play.

7.4.1 A Model for Fictitious Play

Throughout this section, we consider a strategic-form game with K players. It is assumed that K players play the strategic-form game at times $t = 1, \ldots$ Player k has a weight function $w_k^{(t)}(\mathbf{s}_{-k}) : S_{-k} \to \mathbb{R}^+$, where $S_{-k} = \prod_{l \neq k} S_l$, that is updated as follows:

$$w_k^{(t)}(\mathbf{s}_{-k}) = w_k^{(t-1)}(\mathbf{s}_{-k}) + \begin{cases} 1, & \text{if } \mathbf{s}_{-k}^{(t-1)} = \mathbf{s}_{-k}, \\ 0, & \text{otherwise,} \end{cases} \tag{7.35}$$

where $\mathbf{s}_{-k}^{(t-1)}$ represents the actions of the other players at time $t - 1$. The initial weight functions, $w_k^{(0)}(\mathbf{s}_{-k}) \geq 0$, can be given from the past experiences or prior information.

Then, the following empirical distribution can be obtained:

$$p_k^{(t)}(\mathbf{s}_{-k}) = \frac{w_k^{(t)}(\mathbf{s}_{-k})}{\sum_{\mathbf{s}'_{-k} \in S_{-k}} w_k^{(t)}(\mathbf{s}'_{-k})}, \quad \mathbf{s}_{-k} \in S_{-k}. \tag{7.36}$$

Using the empirical distribution of $\mathbf{s}_{-k}$, player k can find an estimate of the payoff as follows:

$$u_k\left(s_k, p_k^{(t)}\right) = \sum_{\mathbf{s}_{-k} \in S_{-k}} u_k(s_k, \mathbf{s}_{-k}) p_k^{(t)}(\mathbf{s}_{-k}), \quad s_k \in S_k. \tag{7.37}$$

From the estimate of the payoff, player k can choose the action to maximize the payoff at time t as follows:

$$s_k^{(t)} \in \mathrm{BR}_k\left(p_k^{(t)}\right), \tag{7.38}$$

where

$$\mathrm{BR}\left(p_k^{(t)}\right) = \underset{s_k \in S_k}{\mathrm{argmax}} \; u_k\left(s_k, p_k^{(t)}\right). \tag{7.39}$$

The approach in (7.38) is called fictitious play as players choose the actions based on the empirical distribution in (7.36), not the true distribution or mixed strategies. This approach is myopic as each player is to maximize the current payoff without taking into account any future payoffs.

In two-person NOMA-ALOHA, each player may learn the other player's action based on the feedback from BS as shown in Table 7.4. For example, if player k chooses H and the feedback is (negative acknowledgment) "NACK," it means that the other player chooses H, leading to a collision. It is noteworthy that the feedback may not be sufficient to identify the other player's action in some cases. For example, if player k chooses L and the feedback is (acknowledgment) "ACK," it implies that the other player may choose either H or 0. To avoid this ambiguity, the BS needs to provide additional information such as the number of received signals (i.e. 1 or 2) when it sends ACK.

Table 7.4 The other player's action based on the feedback from BS.

Player k	NACK	ACK	Idle
H	H	L or 0	Invalid
L	L	H or 0	Invalid
0	Invalid	H or L	0

In the table, ACK and NACK represent positive and negative acknowledgment, respectively.

7.4.2 Convergence

Fictitious play may result in a mixed strategy NE or not depending on conditions. In this section, to see the convergence[3] of fictitious play, we focus on a symmetric game where all players have the same payoff function (which is the case of NOMA-ALOHA).

For fictitious play, from (7.36) (ignoring the prior), we now assume that the empirical distribution is updated as follows:

$$p_k^{(t)} = \frac{t-1}{t} p_k^{(t-1)} + \frac{1}{t} \text{BR}_k\left(p_{-k}^{(t-1)}\right), \tag{7.40}$$

where $p_{-k}^{(t)} = \{p_1^{(t)}, \dots, p_{k-1}^{(t)}, p_{k+1}^{(t)}, \dots, p_K^{(t)}\}$. Let $\tau = \log t$ or $t = \exp(\tau)$ and $\tilde{p}_k^{(\tau)} = p_k^{(\exp(\tau))}$. Then, since $\exp(\tau + \Delta) - \exp(\tau) = \exp(\tau)(\exp(\Delta) - 1) \approx \exp(\tau)\Delta = \Delta t$ (for $\Delta \ll 1$), we have

$$\begin{aligned}
\tilde{p}_k^{(\tau+\Delta)} &= p_k^{(\exp(\tau+\Delta))} = p_k^{(\exp(\tau)+\Delta t)} \\
&= p_k^{(t+\Delta t)} \\
&= \frac{t-\Delta t}{t} p_k^{(t)} + \frac{\Delta t}{t} \text{BR}\left(p_{-k}^{(t-1)}\right) \\
&= (1-\Delta)\tilde{p}_k^{(\tau)} + \Delta\text{BR}\left(\tilde{p}_{-k}^{(\tau)}\right)
\end{aligned} \tag{7.41}$$

under the assumption that play remains more or less constant between τ and $\tau + \Delta$. For large t and small Δ, it can be shown that

$$\frac{d\tilde{p}_k^{(t)}}{dt} = \text{BR}_k\left(\tilde{p}_{-k}^{(t)}\right) - \tilde{p}_k^{(t)}, \tag{7.42}$$

where t is now continuous time, which is referred to as the continuous time fictitious play dynamics.

In a symmetric game, we have

$$u_k(\sigma_k, \sigma_{-k}) = u(\sigma_k, \sigma_{-k}), \quad k = 1, \dots, K.$$

---

3 The proof of convergence is based on the approach in Menache and Ozdaglar (2011).

Then, it can be shown that

$$
\frac{\mathrm{d}}{\mathrm{d}t} u\left(\tilde{p}^{(t)}\right) = \frac{\mathrm{d}}{\mathrm{d}t}\left(\sum_{s_1 \in S} \cdots \sum_{s_K \in S} \tilde{p}_1^{(t)}(s_1) \cdots \tilde{p}_K^{(t)}(s_K) u(s_1, \ldots, s_K)\right)
$$

$$
= \sum_k \sum_{s_1 \in S} \cdots \sum_{s_K \in S} \frac{\mathrm{d}\tilde{p}_k^{(t)}}{\mathrm{d}t}(s_k)\left(\prod_{l \neq k} \tilde{p}_l^{(t)}\right) u(\mathbf{s})
$$

$$
= \sum_k u\left(\frac{\mathrm{d}\tilde{p}_k^{(t)}}{\mathrm{d}t}(s_k), \tilde{p}_{-k}^{(t)}\right). \tag{7.43}
$$

Let

$$
\Delta_k(\sigma_k, \sigma_{-k}) = \max_{\sigma'_k \in \Sigma} u(\sigma'_k, \sigma_{-k}) - u(\sigma_k, \sigma_{-k}). \tag{7.44}
$$

Then, using (7.42), it can be shown that

$$
u\left(\frac{\mathrm{d}\tilde{p}_k^{(t)}}{\mathrm{d}t}(s_k), \tilde{p}_{-k}^{(t)}\right) = u\left(\mathrm{BR}_k\left(\tilde{p}_{-k}^{(t)}\right), \tilde{p}_{-k}^{(t)}\right) - u\left(\tilde{p}_k^{(t)}, \tilde{p}_{-k}^{(t)}\right)
$$

$$
= \Delta_k\left(\tilde{p}_k^{(t)}, \tilde{p}_{-k}^{(t)}\right) = \Delta_k(p^{(t)}) \geq 0, \tag{7.45}
$$

where the inequality is valid due to (7.44). Letting $W(t) = \sum_k \Delta_k(\tilde{p}^{(t)})$, from (7.43) and (7.45), it can be shown that

$$
W(t) = \frac{\mathrm{d}}{\mathrm{d}t} u\left(\tilde{p}^{(t)}\right) \geq 0. \tag{7.46}
$$

Since $W(t)$ is nonnegative, $u(p^{(t)})$ should be nondecreasing as t increases. Furthermore, if $u(\sigma)$ is finite for all σ, it can be shown that $\lim_{t \to \infty} u(\tilde{p}^{(t)}) = u^*$ exists. As a result, we have

$$
u^* - u\left(\tilde{p}^{(t)}\right) \geq u\left(\tilde{p}^{(t+\tau)}\right) - u\left(\tilde{p}^{(t)}\right)
$$

$$
= \int_t^{t+\tau} W(z)\mathrm{d}z
$$

$$
\geq 0, \tag{7.47}
$$

where the first inequality is valid as $u(p^{(t)})$ should be nondecreasing as t increases and the equality is due to (7.46). By the definition of u^*, we see that $u^* - u(\tilde{p}^{(t)}) \to 0$ as $t \to \infty$. Thus, we have $W(t) \to 0$ or $\Delta_k\left(\tilde{p}_k^{(t)}, \tilde{p}_{-k}^{(t)}\right) \to 0$, $k = 1, \ldots, K$, as $t \to \infty$. Finally, it can be shown that

$$
\lim_{t \to \infty} \sum_k \Delta_k\left(\tilde{p}^{(t)}\right) = \lim_{t \to \infty} u\left(\mathrm{BR}_k\left(\tilde{p}_{-k}^{(t)}\right), \tilde{p}_{-k}^{(t)}\right) - u\left(\tilde{p}_k^{(t)}, \tilde{p}_{-k}^{(t)}\right), \tag{7.48}
$$

which implies $\tilde{p}_k^{(t)}$ is asymptotically a best response to $\tilde{p}_{-k}^{(t)}$. In other words, a mixed strategy NE can be obtained using fictitious play with the dynamics in (7.42).

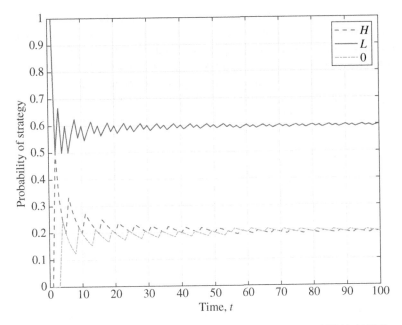

Figure 7.6 Trajectory of the empirical probability of two-person NOMA-ALOHA game using fictitious play with the dynamics in (7.42) when $W = 2.5$.

In Figure 7.6, the trajectory of $p_k^{(t)}$ is shown for two-person NOMA-ALOHA game using fictitious play with the dynamics in (7.42) when $W = 2.5$. According to Figure 7.1, the mixed strategy NE is $\sigma^* = (a^*, b^*, 1 - a^* - b^*) = (0.2, 0.6, 0.2)$. As shown in Figure 7.6, fictitious play can find the mixed strategy NE.

It is noteworthy that fictitious play can converge under specific conditions (e.g. all players have the same payoff as considered above). There are games in which fictitious play does not converge.

7.5 Evolutionary Game Theory and Its Application

In this section, we review evolutionary game theory and consider its application to NOMA so that an optimal power control policy can be derived.

7.5.1 Population Games

Evolutionary games provide a framework to understand strategic interactions in large populations, where each player has the same finite set of strategies, denoted by $\mathcal{L} = \{1, \ldots, L\}$. Here, L stands for the number of strategies. The aggregate behavior of the players is captured by a population state $\mathbf{x} = [x_1 \; \cdots \; x_L]^{\mathrm{T}} \in \mathcal{X}$, where x_l represents the proportion of the players choosing strategy $l \in \mathcal{L}$ and $\mathcal{X}$ is the $(L-1)$-dimensional simplex. Thus, we have $\sum_{l=1}^{L} x_l = 1$ and $x_l \geq 0$. Clearly, $\mathbf{x}$ can also be seen as a mixed strategy adopted by all players or population.

Suppose that a player chooses strategy l when the population state is $\mathbf{x}$ and let $f(l, \mathbf{x})$ denote the payoff or fitness to strategy l. If a player can choose a mixed strategy, the average fitness is given by

$$\bar{f}(\mathbf{y}, \mathbf{x}) = \sum_{l=1}^{L} y_l f(l, \mathbf{x}), \tag{7.49}$$

where $\mathbf{y} = [y_1 \; \cdots \; y_L]^{\mathrm{T}} \in \mathcal{X}$ represents the mixed strategy of the player, when the population state is $\mathbf{x}$. Then, it is possible to define a NE, denoted by $\mathbf{x}^*$. In particular, $\mathbf{x}^*$ is a NE iff

$$\bar{f}(\mathbf{x}^*, \mathbf{x}^*) > \bar{f}(\mathbf{x}, \mathbf{x}^*) \text{ for any } \mathbf{x} \in \mathcal{X} \setminus \mathbf{x}^*. \tag{7.50}$$

Clearly, no player has an incentive to deviate from a NE, $\mathbf{x}^*$.

7.5.2 Replicator Dynamics and Evolutionary Stable State

Suppose that there are n_l players using strategy l at time t and assume that the number n_l of players evolves according to

$$\frac{dn_l}{dt} = n_l f(l, \mathbf{x}), \tag{7.51}$$

where $x_l = \frac{n_l}{\sum_{l=1}^{L} n_l}$. For convenience, let $N = \sum_{l=1}^{L} n_l$. Then, it can be shown that

$$\frac{dx_l}{dt} = \frac{\frac{dn_l}{dt}}{N} - n_l \frac{\sum_{m=1}^{L} \frac{dn_m}{dt}}{N^2}$$

$$= x_l f(l, \mathbf{x}) - x_l \sum_{m=1}^{L} x_m f(m, \mathbf{x})$$

$$= x_l \left(f(l, \mathbf{x}) - \bar{f}(\mathbf{x}, \mathbf{x}) \right), \tag{7.52}$$

which is called the replicator equation. It is readily shown that $\sum_{l=1}^{L} \frac{dx_l}{dt} = \sum_{l=1}^{L} x_l f(l, \mathbf{x}) - \sum_{l=1}^{L} x_l \bar{f}(\mathbf{x}, \mathbf{x}) = 0$. Thus, as long as the initial state is in $\mathcal{X}$, the state at time t is always in the $(L-1)$-dimensional simplex, i.e. $\mathbf{x}(t) \in \mathcal{X}$.

Suppose that $\mathbf{x}^*$ is the mixed strategy adopted by the majority or resident of the population and $\mathbf{x}$ is a mixed strategy by mutants. The proportion of the mutants is denoted by ϵ. Let $\mathbf{y}$ be the state of population in the presence of mutants. Then, the payoffs to strategies $\mathbf{x}$ or $\mathbf{x}^*$ become

$$\bar{f}(\mathbf{x}, \mathbf{y}) = \bar{f}(\mathbf{x}, \mathbf{x}^*)(1 - \epsilon) + \bar{f}(\mathbf{x}, \mathbf{x})\epsilon,$$

$$\bar{f}(\mathbf{x}^*, \mathbf{y}) = \bar{f}(\mathbf{x}^*, \mathbf{x}^*)(1 - \epsilon) + \bar{f}(\mathbf{x}^*, \mathbf{x})\epsilon. \tag{7.53}$$

According to the replicator equation in (7.52), it can be shown that

$$
\begin{aligned}
\frac{d\epsilon}{dt} &= \epsilon \left(\bar{f}(\mathbf{x}, \mathbf{y}) - \left(\epsilon \bar{f}(\mathbf{x}, \mathbf{y}) + (1 - \epsilon) \bar{f}(\mathbf{x}^*, \mathbf{y}) \right) \right) \\
&= \epsilon(1 - \epsilon) \left[(1 - \epsilon) \left(\bar{f}(\mathbf{x}, \mathbf{x}^*) - \bar{f}(\mathbf{x}^*, \mathbf{x}^*) \right) + \epsilon \left(\bar{f}(\mathbf{x}, \mathbf{x}) - \bar{f}(\mathbf{x}^*, \mathbf{x}) \right) \right].
\end{aligned}
\tag{7.54}
$$

Then, the condition that the number of mutants decreases when the majority or resident adopting $\mathbf{x}^*$ is given by

$$(1 - \epsilon) \left(\bar{f}(\mathbf{x}, \mathbf{x}^*) - \bar{f}(\mathbf{x}^*, \mathbf{x}^*) \right) + \epsilon \left(\bar{f}(\mathbf{x}, \mathbf{x}) - \bar{f}(\mathbf{x}^*, \mathbf{x}) \right) < 0, \tag{7.55}$$

which leads to $\frac{d\epsilon}{dt} < 0$ so that $\epsilon(t) \to 0$ (note that $\epsilon(t)$ has to be greater than or equal to 0 as it is the proportion of mutants). From this, the definition of evolutionary stable state (ESS) can be obtained. The mixed strategy $\mathbf{x}^*$ is an ESS if

$$\epsilon \bar{f}(\mathbf{x}^*, \mathbf{x}) + (1 - \epsilon) \bar{f}(\mathbf{x}^*, \mathbf{x}^*) > \epsilon \bar{f}(\mathbf{x}, \mathbf{x}) + (1 - \epsilon) \bar{f}(\mathbf{x}, \mathbf{x}^*), \tag{7.56}$$

for all $\mathbf{x} \neq \mathbf{x}^*$ and $\epsilon < \bar{\epsilon}$, where $\bar{\epsilon} \in (0, 1)$. In other words, as long as the proportion of mutants is less than $\bar{\epsilon}$, the number of mutants decreases, i.e. $\epsilon(t) \to 0$, if (7.56) holds (since (7.56) is equivalent to (7.55)). A sufficient condition for the existence of an ESS can also be found as follows:

$$\bar{f}(\mathbf{x}^*, \mathbf{x}^*) > \bar{f}(\mathbf{x}, \mathbf{x}^*), \quad \text{for all } \mathbf{x} \neq \mathbf{x}^*, \text{ or} \tag{7.57}$$

$$\text{if } \bar{f}(\mathbf{x}^*, \mathbf{x}^*) = \bar{f}(\mathbf{x}, \mathbf{x}^*), \quad \text{then } \bar{f}(\mathbf{x}^*, \mathbf{x}) > \bar{f}(\mathbf{x}, \mathbf{x}). \tag{7.58}$$

The first condition in (7.57) means that $\mathbf{x}^*$ is a NE[4] (thus, it is referred to as the NE condition), while the second condition in (7.58) implies that if the mutant performs as well as the resident against the resident, it does not perform better than the resident against the mutant. The second condition is referred to as the stability condition.

7.5.3 Stability of the Replicator Dynamics

In order to see the reason why (7.58) is the stability condition, consider the following function:

$$V(\mathbf{x}) = \prod_l x_l^{x_l^*}, \tag{7.59}$$

where the product is taken over $\{l : x_l^* > 0\}$. Thus, $V(\mathbf{x}) > 0$ for all $\mathbf{x} \in \mathcal{X}$. Actually, the function $V(\mathbf{x})$ is a Lyapunov function. Lyapunov functions are used to

---

4 As a result, an ESS is a NE (but a NE is not necessarily an ESS).

prove the stability of an equilibrium of an ordinary differential equation (ODE) such as the replicator equation in (7.52).

It can be shown that $V(\mathbf{x})$ is maximized at $\mathbf{x} = \mathbf{x}^*$. We now show that once $\mathbf{x}$ is sufficiently close to $\mathbf{x}^*$ (i.e. an ESS), it will converge to $\mathbf{x}^*$ using $V(\mathbf{x})$. With the replicator equation in (7.52), it can be shown that

$$
\begin{aligned}
\frac{dV}{dt} &= \sum_l x_l^* x_l^{x_l^*-1} \frac{dx_l}{dt} \left(\prod_{l' \neq l} x_{l'}^{x_{l'}^*} \right) \\
&= \sum_l x_l^* x_l^{x_l^*-1} \underbrace{x_l(\bar{f}(l,\mathbf{x}) - \bar{f}(\mathbf{x},\mathbf{x}))}_{= \text{Eq. (7.52)}} \left(\prod_{l' \neq l} x_{l'}^{x_{l'}^*} \right) \\
&= \sum_l x_l^* (\bar{f}(l,\mathbf{x}) - \bar{f}(\mathbf{x},\mathbf{x})) V(\mathbf{x}) \\
&= V(\mathbf{x}) \left(\bar{f}(\mathbf{x}^*,\mathbf{x}) - \bar{f}(\mathbf{x},\mathbf{x}) \right).
\end{aligned}
\tag{7.60}
$$

If $\bar{f}(\mathbf{x}^*,\mathbf{x}^*) = \bar{f}(\mathbf{x},\mathbf{x}^*)$, for an ESS $\mathbf{x}^*$, from (7.58), we have $\bar{f}(\mathbf{x}^*,\mathbf{x}) > \bar{f}(\mathbf{x},\mathbf{x})$, which implies that $\frac{dV}{dt} > 0$ for $\mathbf{x} \neq \mathbf{x}^*$. Otherwise (i.e. $\bar{f}(\mathbf{x}^*,\mathbf{x}^*) \neq \bar{f}(\mathbf{x},\mathbf{x}^*)$), we need to consider (7.57), i.e. $\bar{f}(\mathbf{x}^*,\mathbf{x}^*) > \bar{f}(\mathbf{x},\mathbf{x}^*)$, which is valid for $\mathbf{x} \approx \mathbf{x}^*$ thanks to the continuity of $\bar{f}(\cdot)$, i.e. $\bar{f}(\mathbf{x}^*,\mathbf{x}) > \bar{f}(\mathbf{x},\mathbf{x})$. Thus, as long as $\mathbf{x}$ is sufficiently close to $\mathbf{x}^*$, it converges to the local maximum $\mathbf{x}^*$ as $\frac{dV}{dt} > 0$ (i.e. the replicator dynamics brings $\mathbf{x}$ to $\mathbf{x}^*$), which is locally asymptotically stable.

7.5.4 Application to NOMA

In this section, we consider a game where two users are to transmit their signals on a shared channel with a certain power control policy based on power-domain NOMA. Based on the notion of evolutionary game (although the number of users is two), we find the optimal strategy.

For power-domain NOMA, let γ_k be the channel gain of user $k \in \{1, 2\}$, and the transmit power can be given by

$$
P_k(\gamma_k) = \begin{cases} \frac{P_H}{\gamma_k}, & \text{if } \gamma_k > \tau_{pn}, \\ \frac{P_L}{\gamma_k}, & \text{if } \tau < \gamma_k \leq \tau_{pn}, \\ 0, & \text{if } \gamma_k \leq \tau, \end{cases}
\tag{7.61}
$$

where P_H and P_L are the receive power levels ($P_H > P_L$), and $\tau_{pn} > \tau$. Note that τ_{pn} is another threshold to be determined.

Suppose that there is the strategy set of the three actions. Action 1 is the transmission of high power, i.e. $P_k = \frac{P_H}{\gamma_k}$; action 2 is the transmission of low power, i.e.

$P_k = \frac{P_\perp}{\gamma_k}$; and action 3 is no transmission, i.e. $P_k = 0$. It is noteworthy that since each user's action is decided by γ_k, which is a random variable, a user's selection of strategy can be seen as random to the other user.

Let x_l represent the probability of action $l \in \{1, 2, 3\}$. According to the power control in (7.61), we have

$$x_l = \Pr(\gamma_k \in \mathcal{G}_l), \tag{7.62}$$

where $\mathcal{G}_1 = \{\gamma_k : \gamma_k > \tau_{pn}\}$, $\mathcal{G}_2 = \{\gamma_k : \tau < \gamma_k \leq \tau_{pn}\}$, and $\mathcal{G}_3 = \{\gamma_k : 0 < \gamma_k \leq \tau\}$. The probabilities of actions are dependent on τ and τ_{pn}. In addition, let the set of the probabilities over the actions be $\mathcal{X} = \{\mathbf{x} : \sum_{i=1}^{3} x_l = 1, x_l \geq 0\}$, where $\mathbf{x} = [x_1 \ x_2 \ x_3]^T$ is the probability distribution over three actions (or pure strategies). Note that a distribution $\mathbf{x}$ is also called the state or profile of the population in evolutionary games.

Let us consider the average payoff of user 1, when user 2 employs the state $\mathbf{x}$. The average payoff of user 1 with action $l \in \{1, 2\}$ is given by

$$u_1(l, \mathbf{x}) = R\mathbb{E}_{\mathbf{x}}[\mathbb{1}(\text{succeed with action } l)] - C_l, \tag{7.63}$$

where R is the reward for successful transmission, $\mathbb{E}_{\mathbf{x}}[\cdot]$ is the expectation with respect to the distribution $\mathbf{x}$, and $\mathbb{1}(\cdot)$ represents the indicator function. Here, C_l represents the cost of action l. In addition, we have

$$u_1(3, \mathbf{x}) = -C_3. \tag{7.64}$$

Let $\bar{\mathbf{x}} = [\bar{x}_1 \ \bar{x}_2 \ \bar{x}_3]^T \in \mathcal{X}$. Then, the average payoff of user 1 with a state (mixed strategy) $\bar{\mathbf{x}}$ becomes

$$u(\bar{\mathbf{x}}, \mathbf{x}) = \sum_{l=1}^{3} \bar{x}_l u_1(l, \mathbf{x}) = \bar{\mathbf{x}}^T \mathbf{A} \mathbf{x}, \tag{7.65}$$

where

$$\mathbf{A} = \begin{bmatrix} -C_1 & R - C_1 & R - C_1 \\ R - C_2 & -C_2 & R - C_2 \\ -C_3 & -C_3 & -C_3 \end{bmatrix}. \tag{7.66}$$

This shows that it is a symmetric game.

Let $f(l, \mathbf{x}) = u_1(l, \mathbf{x})$ and $\bar{f}(\bar{\mathbf{x}}, \mathbf{x}) = u(\bar{\mathbf{x}}, \mathbf{x})$. Then, using the replicator dynamics, we are able to find an ESS, which is also a NE. In particular, the following replicator equation[5] can be considered:

$$\dot{x}_l = \mu x_l(f(l, \mathbf{x}) - \bar{f}(\mathbf{x}, \mathbf{x})), \tag{7.67}$$

where $\mu > 0$ is the step-size.

---

5 In the replicator equation, $\dot{x}_l = \frac{d}{dt} x_l(t)$, where $x_l(t)$ represents x_l at time t. In a discrete-time system, $\dot{x}_l = x_l(t + 1) - x_l(t)$, where $t \in \mathbb{Z}$ represents the discrete time unit.

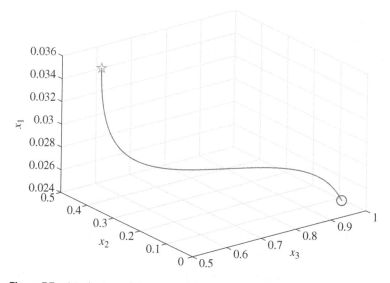

Figure 7.7 A trajectory of the state of the replicator equation in (7.67) with the initial state of $(x_1, x_2, x_3) = (0.025, 0.025, 0.95)$ that is represented by ∘ marker, when $R = 1$, $c = 2, \Gamma = 4$ (or 6 dB), $\bar{\gamma} = 10$ (or 10 dB), $\mu = 0.2$. The replicator dynamics converges to $(x_1^*, x_2^*, x_3^*) = (0.035, 0.415, 0.550)$, which is represented by ★ marker.

For example, suppose that the cost is proportional to the average transmit power, i.e.

$$C_l = \mathbb{E}\left[cP_k(\gamma_k) \mid \gamma_k \in \mathcal{G}_l\right], \tag{7.68}$$

and $\gamma_k \sim Exp(\bar{\gamma})$, where $\bar{\gamma}$ is the mean channel gain. Let $P_{\rm H} = \Gamma(1 + \Gamma)$ and $P_{\rm L} = \Gamma$. With $(R, c) = (1, 2)$ and $(\Gamma, \bar{\gamma}) = (6$ dB, 10 dB), from (7.65), there is a unique ESS, which can be analytically found as

$$(x_1^*, x_2^*, x_3^*) = (0.035, 0.415, 0.550),$$

which can also be found by the replicator equation. In Figure 7.7, with $\mu = 0.2$, the trajectory of $\mathbf{x}(t)$ is illustrated according to (7.67) to find the ESS, where we see that $\mathbf{x}(t)$ converges to $\mathbf{x}^*$.

7.6 Further Reading

There are a number of textbooks on game theory, e.g. Fudenberg and Tirole (1991), Osborne and Rubinstein (1994), and Maschler et al. (2013). Applications of game theory to communications and networks can be found in Scutari et al. (2010), Chen et al. (2010), Menache and Ozdaglar (2011), and Bacci et al. (2016). To see the details of NOMA-ALOHA game, the reader is referred to Choi (2018a).

In this chapter, we also discussed fictitious play that is related to learning. The reader is referred to Fudenberg and Levine (1998) to see learning aspects in games in details.

Bibliography

3GPP (2016). *Evolved Universal Terrestrial Radio Access (E-UTRA); Medium Access Control (MAC) protocol specification*. 3GPP TS 36.321 V13.2.0, June 2016.

3GPP (2018). *Evolved Universal Terrestrial Radio Access (EUTRA) and Evolved Universal Terrestrial Radio Access Network (EUTRAN); Overall Description, TS 36.300 v.14.7.0*. 3rd Generation Partnership Project (3GPP), June 2018.

Alouini, M.-S. and Goldsmith, A.J. (1999). Capacity of Rayleigh fading channels under different adaptive transmission and diversity-combining techniques. *IEEE Trans. Veh. Technol.* 48 (4): 1165–1181.

Applebaum, L., Bajwa, W.U., Duarte, M.F., and Calderbank, R. (2012). Asynchronous code-division random access using convex optimization. *Phys. Commun.* 5 (2): 129–147.

Bacci, G., Lasaulce, S., Saad, W., and Sanguinetti, L. (2016). Game theory for networks: a tutorial on game-theoretic tools for emerging signal processing applications. *IEEE Signal Process. Mag.* 33 (1): 94–119.

Bertsekas, B. and Gallager, R. (1987). *Data Networks*. Englewood Cliffs, NJ: Prentice-Hall.

Biglieri, E. (2005). *Coding for Wireless Channels*. New York: Springer.

Björnson, E., de Carvalho, E., Sørensen, J.H. et al. (2017). A random access protocol for pilot allocation in crowded massive MIMO systems. *IEEE Trans. Wireless Commun.* 16 (4): 2220–2234. https://doi.org/10.1109/TWC.2017.2660489.

Björnson, E., Hoydis, J., and Sanguinetti, L. (2018). Massive MIMO has unlimited capacity. *IEEE Trans. Wireless Commun.* 17 (1): 574–590. https://doi.org/10.1109/TWC.2017.2768423.

Blei, D.M., Kucukelbir, A., and McAuliffe, J.D. (2017). Variational inference: a review for statisticians. *J. Am. Stat. Assoc.* 112 (518): 859–877.

Bockelmann, C., Pratas, N., Nikopour, H. et al. (2016). Massive machine-type communications in 5G: physical and MAC-layer solutions. *IEEE Commun. Mag.* 54 (9): 59–65.

Massive Connectivity: Non-Orthogonal Multiple Access to High Performance Random Access,
First Edition. Jinho Choi.
© 2022 The Institute of Electrical and Electronics Engineers, Inc. Published 2022 by John Wiley & Sons, Inc.

Bonnefoi, R., Besson, L., Moy, C. et al. (2018). Multi-armed bandit learning in IoT networks: learning helps even in non-stationary settings. In: *Cognitive Radio Oriented Wireless Networks*, (ed. P. Marques, A. Radwan, S. Mumtaz et al.), 173–185. Cham: Springer International Publishing. ISBN 978-3-319-76207-4.

Boyd, S. and Vandenberghe, L. (2009). *Convex Optimization*. Cambridge University Press.

Candes, E.J., Romberg, J., and Tao, T. (2006). Robust uncertainty principles: exact signal reconstruction from highly incomplete frequency information. *IEEE Trans. Inf. Theory* 52 (2): 489–509.

Casini, E., De Gaudenzi, R., and Rio Herrero, O.D. (2007). Contention resolution diversity slotted ALOHA (CRDSA): an enhanced random access scheme for satellite access packet networks. *IEEE Trans. Wireless Commun.* 6 (4): 1408–1419. https://doi.org/10.1109/TWC.2007.348337.

Chen, J. and Huo, X. (2006). Theoretical results on sparse representations of multiple-measurement vectors. *IEEE Trans. Signal Process.* 54 (12): 4634–4643.

Chen, L., Low, S.H., and Doyle, J.C. (2010). Random access game and medium access control design. *IEEE/ACM Trans. Networking* 18 (4): 1303–1316.

Choi, J. (2016). Re-transmission diversity multiple access based on SIC and HARQ-IR. *IEEE Trans. Commun.* 64 (11): 4695–4705.

Choi, J. (2017a). Joint rate and power allocation for NOMA with statistical CSI. *IEEE Trans. Commun.* 65 (10): 4519–4528.

Choi, J. (2017b). NOMA-based random access with multichannel ALOHA. *IEEE J. Sel. Areas Commun.* 35 (12): 2736–2743.

Choi, J. (2018a). Multichannel NOMA-ALOHA game with fading. *IEEE Trans. Commun.* 66 (10): 4997–5007.

Choi, J. (2018b). Stability and throughput of random access with CS-based MUD for MTC. *IEEE Trans. Veh. Technol.* 67 (3): 2607–2616. ISSN 0018-9545.

Choi, J. (2019a). Low-latency multichannel ALOHA with fast retrial for machine-type communications. *IEEE Internet Things J.* 6 (2): 3175–3185.

Choi, J. (2019b). NOMA-based compressive random access using Gaussian spreading. *IEEE Trans. Commun.* 67 (7): 5167–5177.

Choi, J. (2010). *Optimal Combining and Detection*. Cambridge University Press.

Choi, J. (2021). An approach to preamble collision reduction in grant-free random access with massive MIMO. *IEEE Trans. Wireless Commun.* 20 (3): 1557–1566. https://doi.org/10.1109/TWC.2020.3034308.

Choi, J. and Ding, J. (2021). Co-existing preamble and data transmissions in random access for MTC with massive MIMO. *IEEE Trans. Commun.* 1. https://doi.org/10.1109/TCOMM.2021.3105693.

Choi, J., Ding, J., Le, N.-P., and Ding, Z. (2021). Grant-free random access in machine-type communication: approaches and challenges. *IEEE Wireless Commun.* 1–8. https://doi.org/10.1109/MWC.121.2100135.

Choi, Y.-J., Park, S., and Bahk, S. (2006). Multichannel random access in OFDMA wireless networks. *IEEE J. Sel. Areas Commun.* 24 (3): 603–613.

Chu, D. (1972). Polyphase codes with good periodic correlation properties (Corresp.). *IEEE Trans. Inf. Theory* 18 (4): 531–532.

Cover, T.M. and Thomas, J.A. (2006). *Elements of Information Theory*, 2e. Hoboken, New Jersey: Wiley.

Dai, L., Wang, B., Yuan, Y. et al. (2015). Non-orthogonal multiple access for 5G: solutions, challenges, opportunities, and future research trends. *IEEE Commun. Mag.* 53 (9): 74–71.

Dai, L., Wang, B., Ding, Z., Wang, Z. et al. (2018). A survey of non-orthogonal multiple access for 5G. *IEEE Commun. Surv. Tutorials* 20 (3): 2294–2323.

Davies, M.E. and Eldar, Y.C. (2012). Rank awareness in joint sparse recovery. *IEEE Trans. Inf. Theory* 58 (2): 1135–1146.

Ding, Z., Liu, Y., Choi, J. et al. (2017). Application of non-orthogonal multiple access in LTE and 5G networks. *IEEE Commun. Mag.* 55 (2): 185–191.

Ding, J., Nemati, M., Ranaweera, C., and Choi, J. (2020a). IoT connectivity technologies and applications: a survey. *IEEE Access* 8: 67646–67673.

Ding, J., Qu, D., and Choi, J. (2020b). Analysis of non-orthogonal sequences for grant-free RA with massive MIMO. *IEEE Trans. Commun.* 68 (1): 150–160. https://doi.org/10.1109/TCOMM.2019.2950014.

Donoho, D.L. (2006). Compressed sensing. *IEEE Trans. Inf. Theory* 52 (4): 1289–1306.

Eldar, Y.C. and Kutyniok, G. (2012). *Compressed Sensing: Theory and Applications*. Cambridge University Press.

Felegyhazi, M. and Hubaux, J.-P. (2006). Game Theory in Wireless Networks: A Tutorial. *Technical report*, EPFL.

Foucart, S. and Rauhut, H. (2013). *A Mathematical Introduction to Compressive Sensing*. Springer.

Fudenberg, D. and Levine, D.K. (1998). *The Theory of Learning in Games*. Cambridge, MA: MIT Press.

Fudenberg, D. and Tirole, J. (1991). *Game Theory*. Cambridge, MA: MIT Press.

Gallager, R.G. (2008). *Principles of Digital Communication*. Cambridge University Press. https://doi.org/10.1017/CBO9780511813498.

Ghez, S., Verdu, S., and Schwartz, S.C. (1989). Optimal decentralized control in the random access multipacket channel. *IEEE Trans. Autom. Control* 34 (11): 1153–1164.

Gubbi, J., Buyya, R., Marusic, S., and Palaniswami, M. (2013). Internet of Things (IoT): a vision, architectural elements, and future directions. *Future Gener. Comput. Syst.* 29 (7): 1645–1660. https://doi.org/10.1016/j.future.2013.01.010.

Hardy, G., Littlewood, J.E., and Polya, G. (1952). *Inequalities*, 2e. Cambridge University Press.

Higuchi, K. and Kishiyama, Y. (2013). Non-orthogonal access with random beamforming and intra-beam SIC for cellular MIMO downlink. *IEEE Vehicular Technology Conference (VTC Fall)*, pp. 1–5, September 2013.

Kelly, F. and Yudovina, E. (2014). *Stochastic Networks*. Cambridge University Press.

Kim, B., Lim, S., Kim, H. et al. (2013). Non-orthogonal multiple access in a downlink multiuser beamforming system. *MILCOM 2013 - 2013 IEEE Military Communications Conference*, pp. 1278–1283, November 2013.

Kim, J., Yun, J., Choi, S. et al. (2016). Standard-based IoT platforms interworking: implementation, experiences, and lessons learned. *IEEE Commun. Mag.* 54 (7): 48–54.

Kim, J., Lee, G., Kim, S. et al. (2021). Two-step random access for 5G system: latest trends and challenges. *IEEE Network* 35 (1): 273–279. https://doi.org/10.1109/MNET.011.2000317.

Koutsoupias, E. and Papadimitriou, C. (1999). *Worst-Case Equilibria*, 404–413. Berlin, Heidelberg: Springer-Verlag. https://doi.org/10.1007/3-540-49116-3_38.

Lin, S. and Costello, D.J. Jr. (1983). *Error Control Coding: Fundamentals and Applications*, Englewood Cliffs, NJ: Prentice Hall.

Liva, G. (2011). Graph-based analysis and optimization of contention resolution diversity slotted ALOHA. *IEEE Trans. Commun.* 59 (2): 477–487. https://doi.org/10.1109/TCOMM.2010.120710.100054.

Lu, L., Li, G.Y., Swindlehurst, A.L. et al. (2014). An overview of massive MIMO: benefits and challenges. *IEEE J. Sel. Top. Signal Process.* 8 (5): 742–758. https://doi.org/10.1109/JSTSP.2014.2317671.

Marzetta, T.L. (2010). Noncooperative cellular wireless with unlimited numbers of base station antennas. *IEEE Trans. Wireless Commun.* 9 (11): 3590–3600.

Maschler, M., Zamir, S., and Solan, E. (2013). *Game Theory*. Cambridge University Press.

Menache, I. and Ozdaglar, A. (2011). Network games: theory, models, and dynamics. *Synth. Lect. Commun. Networks* 4 (1): 1–159. https://doi.org/10.2200/s00330ed1v01y201101cnt009.

Metzner, J. (1976). On improving utilization in ALOHA networks. *IEEE Trans. Commun.* 24 (4): 447–448. https://doi.org/10.1109/TCOM.1976.1093317.

Mitzenmacher, M. and Upfal, E. (2005). *Probability and Computing: Randomized Algorithms and Probability Analysis*. Cambridge University Press.

Mutairi, A., Roy, S., and Hwang, G. (2013). Delay analysis of OFDMA-aloha. *IEEE Trans. Wireless Commun.* 12 (1): 89–99. https://doi.org/10.1109/TWC.2012.113012.111776.

Namislo, C. (1984). Analysis of mobile radio slotted aloha networks. *IEEE Trans. Veh. Technol.* 33 (3): 199–204. https://doi.org/10.1109/T-VT.1984.24006.

Nisan, N., Roughgarden, T., Tardos, E., and Vazirani, V.V. (2007). *Algorithmic Game Theory*. New York: Cambridge University Press. ISBN 0521872820.

Norris, J.R. (1998). *Markov Chains, Cambridge Series in Statistical and Probabilistic Mathematics*. Cambridge University Press. ISBN 9780521633963. https://books .google.com.au/books?id=qM65VRmOJZAC.

Osborne, M.J. and Rubinstein, A. (1994). *A Course in Game Theory*, vol. 1. The MIT Press.

Papoulis, A. and Unnikrishna Pillai, S. (2002). *Probability, Random Variables and Stochastic Processes*, 4e. McGraw-Hill.

Proakis, J. (2000). *Digital Communications*, 4e. McGraw-Hill.

Roberts, L.G. (1975). ALOHA packet system with and without slots and capture. *SIGCOMM Comput. Commun. Rev.* 5 (2): 28–42. https://doi.org/10.1145/1024916 .1024920.

Ross, S.M. (1995). *Stochastic Processes, Wiley Series in Probability and Mathematical Statistics*. Wiley. ISBN 9780471120629. https://books.google.com.au/books? id=qiLdCQAAQBAJ.

Saraydar, C.U., Mandayam, N.B., and Goodman, D.J. (2002). Efficient power control via pricing in wireless data networks. *IEEE Trans. Commun.* 50 (2): 291–303.

Scutari, G., Palomar, D.P., Facchinei, F., and Pang, Js. (2010). Convex optimization, game theory, and variational inequality theory. *IEEE Signal Process. Mag.* 27 (3): 35–49.

Tse, D. and Viswanath, P. (2005). *Fundamentals of Wireless Communication*. Cambridge University Press.

Verdu, S. (1998). *Multiuser Detection*. Cambridge University Press.

Viswanath, P., Tse, D.N.C., and Laroia, R. (2002). Opportunistic beamforming using dumb antennas. *IEEE Trans. Inf. Theory* 48 (6): 1277–1294.

Wicker, S.B. (1995). *Error Control Systems for Digital Communication and Storage*, Prentice Hall.

Wunder, G., Jung, P., and Ramadan, M. (2015). Compressive random access using a common overloaded control channel. *2015 IEEE Globecom Workshops (GC Wkshps)*, pp. 1–6. https://doi.org/10.1109/GLOCOMW.2015.7414186.

Yu, Y. and Giannakis, G.B. (2007). High-throughput random access using successive interference cancellation in a tree algorithm. *IEEE Trans. Inf. Theory* 53 (12): 2253–2256.

Zhu, H. and Giannakis, G.B. (2011). Exploiting sparse user activity in multiuser detection. *IEEE Trans. Commun.* 59 (2): 454–465.

Index

Massive Connectivity: Non-Orthogonal Multiple Access to High Performance Random Access,
First Edition. Jinho Choi.
© 2022 The Institute of Electrical and Electronics Engineers, Inc. Published 2022 by John Wiley & Sons, Inc.

Printed and bound by CPI Group (UK) Ltd, Croydon, CR0 4YY
16/08/2022
03141869-0001